FUSION OF HARD AND SOFT CONTROL STRATEGIES FOR THE ROBOTIC HAND

FUSION OF HARD AND SOFT CONTROL STRATEGIES FOR THE ROBOTIC HAND

Cheng-Hung Chen
Desineni Subbaram Naidu

IEEE PRESS

WILEY

Published by John Wiley & Sons, Inc., Hoboken, New Jersey.
Published simultaneously in Canada.

For general information on our other products and services or for technical support, please contact our Customer Care Department within the United States at (800) 762-2974, outside the United States at (317) 572-3993 or fax (317) 572-4002.

Wiley also publishes its books in a variety of electronic formats. Some content that appears in print may not be available in electronic formats. For more information about Wiley products, visit our web site at www.wiley.com.

Library of Congress Cataloging-in-Publication Data is available.

ISBN: 978-1-119-27359-2

Printed in the United States of America.

10 9 8 7 6 5 4 3 2 1

To God

– Cheng-Hung Chen

To Amputees of the World

– Desineni Subbaram Naidu

"Fifty years out, I think, we will have largely eliminated disability."

– Eliza Strickland, IEEE Spectrum, June 2014

CONTENTS

LIST OF FIGURES

LIST OF TABLES

CHAPTER 1

INTRODUCTION

A hand is considered as an agent of human brain and is the most intriguing and versatile appendage to the human body. Over the last several years, attempts were made to build a prosthetic/robotic hand to replace a human hand to fully simulate the various natural/human-like operations of moving, grasping, lifting, twisting, and so on. Replicating the human hand in all its various functions is still a challenging task due to the extreme complexity of a human hand, which has 27 bones, controlled by about 38 muscles to provide the hand with 22 degrees of freedom (DOFs), and incorporates about 17,000 tactile units of four different types [1, 2]. Parallels between dextrous robot and human hands were explored by examining sensor motor integration in the design and control of these robots through bringing together experimental psychologists, kinesiologists, computer scientists, and electrical and mechanical engineers.

In this chapter, we present introductory material on relevance to military, overview of control strategies, fusion of hard and soft control strategies, and summary of the remaining chapters.

Fusion of Hard and Soft Control Strategies for the Robotic Hand, By C.-H. Chen and D. S. Naidu
© 2017 by the Institute of Electrical and Electronic Engineers, Inc. Published 2017 by John Wiley & Sons, Inc.

Background

The proposed book is an outgrowth of the interdisciplinary Biomedical Sciences and Engineering (BMSE) research project exemplifying *The Third Revolution: The Convergence of Life Sciences, Physical Sciences, and Engineering*[1] [3–6]. It is to be noted that the book *Fusion of Hard and Soft Control Strategies for the Robotic Hand* basically focuses on the robotic hand applicable to prosthetic/robotic and non-prosthetic applications starting from industrial [7], operation in chemical and nuclear hazardous environments [8, 9], space station building, repair and maintenance [10, 11], explosive and terrorist situations [12] to robotic surgery [13].

1.1 Relevance to Military

During the recent wars in Afghanistan and Iraq, "at least 251,102 people have been killed and 532,715 people have been seriously wounded" [14]. Further, in the United States, the Amputee Coalition of America (ACA) [15] reports that there are approximately 1.9 million people living with limb loss, due to combat operations (such as conflicts and wars), and non-combat operations such as accidents, or birth defects. According to a study of the 1996 National Health Interview Survey (NHIS) published by Vital and Health Statistics [16], it is estimated that one out of every 200 people in the United States has had an amputation. That is, one in every 2,000 new born babies will have limb deficiency and over 3,000 people lose a limb every week in America. By the year 2050, the projected number of Americans living with limb amputation will become 3.6 million [17].

The following documents reveal the intense interest by military in the area of smart prosthetic/robotic hand.

1. First, according to [18], recognizing that "arm amputees rely on old devices" and that the existing technology for arm and hand amputees was not changed significantly in the past six decades, the Defense Department is embarking on a research program to "fund prosthetics research" according to [19] to revolutionalize upper-body prosthetics and to develop artificial arms that will "feel, look and perform" like a real arm guided by the central nervous system.

2. According to [20, 21], Bio-Revolution is one of the eight strategic research thrusts that DARPA is emphasizing in response to emerging trends and national security. In particular, the Human Assisted Neural Devices program under Bio-Revolution will have "immediate benefit to injured veterans, who would be able to control prosthetics..." A related area of interest in Bio-Revolution is Cell and Tissue Engineering.

3. Next, according to Defense Science Office (DFO) of DARPA [22], emerging technologies for combat casualties care with dual usage for both military and

---

[1] The First Revolution: Molecular and Cellular Biology and The Second Revolution: Genomics

civilian medical care, focus on programs in Revolutionizing Prosthetics, Human Assisted Neural Devices, Biologically Inspired Multi-functional Dynamic Robotics, and so on. In particular, according to [23], "today on of the most devastating battlefield injuries is loss of a limb... at DARPA, the vision of a future is to ... regain full use of that limb again..."

According to an article that appeared in IEEE Spectrum issue of June 2014, "Fifty years out, I think we will have largely eliminated disability" — Eliza Strickland [24]. The robotic hand, in addition to using it for prosthetic applications, is highly useful for performing various operations that a real human hand cannot perform without reaching a fatigue stage and especially for handling of hazardous waste materials and conditions.

Finally, an IEEE video on overview of how engineers are solutionists, poses "What if prosthetics were stronger and more accurate than the human body?" [25]

1.2 Control Strategies

1.2.1 Prosthetic/Robotic Hands

Artificial hands have been around for several years and have been developed by various researchers in the field and some of the prosthetic/robotic devices developed are given below (in chronological order) [2, 26].

1. Russian arm – [27–29]

2. Waseda hand – [30]

3. Boston arm[2] – [31]

4. UNB hand (University of New Brunswick) – [32–34]

5. Hanafusa hand – [35]

6. Crossley hand – [36]

7. Okada hand – [37]

8. Utah/MIT hand (University of Utah/Massachusetts Institute of Technology) – [38–40]

9. JPL/Stanford hand (Jet Propulsion Laboratory/Stanford University) – [41, 42]

10. Minnesota hand – [43]

11. Manus hand – [44, 45]

---

[2]The "Boston Arm," project involved the Harvard Medical School, Massachusetts General Hospital, the Liberty Mutual Research and Rehabilitation Centers, and MIT

12. Kobayashi hand – [46]

13. Rovetta hand – [47]

14. UT/RAL hand – [48]

15. Dextrous gripper – [49]

16. Belgrade/USC hand (University of Belgrade/University of Southern California) – [50]

17. Southampton hand (University of Southampton, Southampton, UK) – [51]

18. MARCUS hand (Manipulation And Reaction Control under User Supervision) – [52]

19. Kobe hand (Kobe University, Japan) – [53]

20. Robonaut hand (NASA Johnson Space Center) – [54]

21. NTU hand (National Taiwan University) – [55]

22. Hokkaido hand – [56]

23. DLR hand (Deutschen Zentrums für Luft- und Raumfahrt-German Aerospace Center) – [57, 58]

24. TUAT/Karlsruhe hand (Tokyo University of Agriculture and Technology/University of Karlsruhe) – [59]

25. BUAA hand (Beijing University of Aeronautics and Astronautics) – [60]

26. TBM hand (Toronto/Bloorview MacMillan) – [61]

27. ULRG System (University of Louisiana Robotic Gripper) – [62]

28. Oxford hand – [44]

29. IOWA hand (University of Iowa) – [63]

30. MA-I hand – [64]

31. RCH-1 (ROBO CASA hand 1[3]) – [65]

32. UB hand (University of Bologna) – [66]

33. Ottobock SUVA hand – (www.ottobock.co.uk)

34. Northwestern University system – [67]

35. SKKU Hand II (Sungkyunkwan University, Korea) – [68]

---

[3] The Italy–Japan joint laboratory for Research on Humanoid and Personal Robotics

36. Applied Physics Laboratory (APL) at Johns Hopkins University (JHU) – [23, 69, 70]

and some of the commercial web sites for prosthetic/robotic devices are

1. Sensor Hand^{TM} Speed from Ottobock (www.ottobock.co.uk),

2. VASI (Variety Ability Systems Inc.), a company of the Otto Bock Group (http://www.vasi.on.ca/index.html),

3. Utah Arm from Motion Control (www.utaharm.com),

4. The i-LIMB Hand from Touch Bionics (www.touchbionics.com), and so on.

A very useful comparison table between several hands listed above and human hand, adapted from [2, 26], is updated and shown in Table 1.1.

However, about 35% of the amputees do not use their prosthetic/robotic hand regularly according to [71] due to various reasons such as poor functionality of the presently available prosthetic/robotic hands and psychological problems. To overcome this problem, one has to design and develop an artificial hand which "mimics the human hand as closely as possible" both in functionality and appearance.

There are a number of surveys, and/or state-of-the-art articles that appeared over the years on the subject of myoelectric prosthetic/robotic hand including the work in USSR (Russian) given by [28] and some of them are given by references [2, 72–84].

1.2.2 Chronological Overview

An overview of the literature on *prosthetic/robotic hand technology*, conducted by the authors [85, 86] is briefly summarized in the next Section 1.2.3. This overview, focusing on recent developments and continuously being updated, is intended to supplement the already existing excellent survey articles [2, 79, 81, 87, 88]. Further, this overview is not intended to be an exhaustive survey on this topic, and any omissions of other works are purely unintentional.

Up to 1970

Electromyographic (EMG) signal is a simple and easily obtainable source of information about the various movements to be used for artificial/prosthetic hands. The EMG extraction using surface electrodes is a very attractive method from the point of view of the user compared to implants requiring surgery. Research activity in the field of prosthetic/robotic limbs was initiated by United States National Academy of Sciences in response to the needs of a large number of casualties in World War II [89]. It was first proposed by [90, 91] the concept of EMG signals for the control of a prosthetic/robotic hand for amputees. A proportional (open-loop) control system, in which the amplitude of the hand motor voltage and hence its speed and force measured from strain gauges varies in direct proportion (linearly) to the amplitude of the EMG signal generated by the prosthetic/robotic hand, was first reported by [92, 93]. In addition, the system added force and velocity feedback controls, so the

Table 1.1   Comparison of Human Hand with Artificial Hands: Robotic and Prosthetic/Robotic Hands: *Force* Indicates Power Grasp *Speed* Indicates the Time Required for a Full Closing and Opening; *E*: Stands for *External*; *I*: Stands for *Internal*

	Size (Norm)	No. of Fingers	No. of DOFs	No. of Sensors	No. of Actuators	Weight (gms)	Force (N)	Speed (sec)	Controls
Human hand	1.0	5	22	≈ 17,000	38(I+E)	≈ 400	>300	0.25	E
Russian arm		5		3	1		≈ 147		
Waseda hand									
UNB hand									
Hanafusa hand									
Crossley hand									
Utah/MIT hand	≈ 2.0	4	16	16	32(E)	-	31.8	-	E
JPL/Stanford hand	≈ 1.2	3	9	-	12(E)	1100	45	-	E
Minnesota hand									
Kobayashi hand									
Rovetta hand									
UT/RAL hand									
Dextrous gripper									
Belgrade/USC hand	≈ 1.1	4	4	23+4	4(E)	-	-	-	E
Southampton hand	≈ 1.0	5	6	-	6(E)	400	38	≈ 5	E
MARCUS hand	≈ 1.1	3	2	3	2(I)	-	-	-	I
Kobe hand									
Robonaut hand	≈ 1.5	5	12+2	43+	14(E)	-	-	-	E
NTU hand	≈ 1	5	17	35	17(E)	1570	-	-	E
Hokkaido hand	≈ 1	5	17	35	17(E)	1570	-	-	E
DLR hand II	>1	5	7	-	7(E)	125	-	-	E
TUAT/Karlsruhe hand	≈ 1	5	17	-	17(E)	≈ 120	12	0.1	E
BUAA hand		4			2				
TBM hand									
Oxford hand									
IOWA hand									
MA-I hand									
Robo Casa hand-1	≈ 1	5	16	24	6(E+I)	350	≈ 40	0.25	E
Ottobock SUVA	1	3	1	2	1(E)	600	-	-	I
UB hand									
Hokkaido hand	>1	5	7	-	7(E)	125	-	-	E
Northwestern									
SKKU Hand II	1.1	4	4	-	3	900			
APL-JHU System									

users could feel more natural to utilize this device. An adaptive control scheme was developed by [94] for a Southampton Hand.

1971–1979

The work reported by [32] studied the effect of sensory feedback based on semi-conductor strain gauges on either edge of thumb of the prosthetic/robotic hand to adjust the stimulus magnitude to target value and avoid dropping or crushing objects for control of a prosthesis and found this acceptable for patients. When the strain gauges received the stimulus, the system amplified and transferred the signals to comparator, and then the comparator modified the range of amplitude of stimulus to the level that the users needed. However, the device with feedback is two or three times larger than the normal hand. A hierarchical method consisting of analytical control theory such as performance-adaptive self-organizing control algorithm and artificial intelligence using fuzzy automaton was presented by [95] to drive a prosthetic/robotic hand.

1980–1989

In providing a historical perspective, the contribution by [72] presented the status of the closed-loop (feedback) control principles for the application of prosthetic/robotic devices, three concepts relating to supplemental sensory feedback, artificial reflexes, and feedback through control interfaces were discussed and it was concluded that "we have not moved very far in the last 65 years in the clinical application of these concepts." A statistical analysis involving the study of zero crossings, second to fifth moments, and correlation functions and pattern classification of EMG signals was given by [96]. A probabilistic model of the EMG pattern was formulated in the feature space of integral absolute value (IAV) to provide the relation between a command, represented by motion and speed variables, and the location and shape of the pattern for real-time control of a prosthetic/robotic arm as given in [97]. Using kinematic relationships for dynamic model of fingers, multi-variable feedback control strategies using pole assignment in frequency domain were employed by [42] to guarantee local stability for controlling one finger of the JPL/Stanford hand. The work in [42] produced the dynamic models of three fingers (thumb, index, and middle) and three joints first, and then used Laplace transform to work in frequency domain. To get a guaranteed stability of control system, the roots/poles had to be located in the left half plane. Hence, they could get a desired steady movement of fingers by controlling the positions of the roots. The works reported by the group [98, 99] were one of the first groups who investigated various aspects such as kinematics, prehensility, dynamics, and control of multi-fingered hands manipulating objects of arbitrary shape in three dimensions.

1990–1999

Design, implementation, and experimental verification of an improved cybernetic elbow prosthesis was presented in [100, 101] that mimics the natural limb to both internal (voluntary) inputs from the amputee and external inputs from the environment. The work in [102] considered a dextrous hand employing a systematic approach to achieve the object stiffness control by actuator position control, tendon tension control, joint torque control, joint stiffness control, and Cartesian fingertip stiffness control. The work by [75] conducted a survey of 33 patients wearing the proportional myoelectric hand grouped into three categories based on previous experience with a terminal device: digital (on–off) myoelectric hand, body-powered terminal device, and no terminal device. The survey resulted in that the group of patients having experience with digital hand "were most impressed with proportionally controlled hand," because it has the advantages: comfortable, cosmetic acceptance, more natural, superior pinch force (11–25 lb) compared to voluntary opening (7–8 lb), a greater range of function but less energy, sensory feedback, force feedback, and short below-elbow.

The research work in [103] developed three tests for evaluation of input–output properties of patient control of neuroprosthetic hand grasp, which compensates or enriches the function of a damaged peripheral nervous system: first test for static input–output properties of the hand grasp, second one for control of hand grasp outputs while tracking step and ramp functions, and finally to obtain the input–output frequency response of the hand grasp system dynamics to estimate the transfer function using spectral analysis. Each test used visual feedback when the users controlled the grasp force and grasp position tracking of the hand. It was shown in [104] that the myoelectric signal (MES) is not random during the initial phase of muscle contraction thus providing a means of classifying patterns from different contraction types. The means is to establish the 60 records of an isometric contraction of the subjects and then produce some anisometric contraction types, like flexion and extension. This information was useful in designing a new multi-function myoelectric control system using artificial neural networks (ANNs) for classifying myoelectric patterns. Additionally, the hidden layer size, segment length, and EMG electrode positions were studied. See related works in [105–108] on multi-functional myoelectric control systems using pattern recognition methods for MES extraction and classification. The control philosophy of a multi-fingered robotic hands for possible adaptation and use in prosthetics and rehabilitation was discussed by [109–111] with respect to the Belgrade/USC robot hand by [50], called PRESHAPE (Programmable Robotic Experimental System for Hands and Prosthetics Evaluation), which estimates a system that translates task commands to motor commands using pressure sensors, force sensors, and pressure feedback which is very useful to detect small contact forces.

Using the dynamic model of the nonlinear neuromuscular (motor servo) control system of human finger muscles including mechanical properties (such as viscoelasticity) of the muscle and stretch reflex, a surface-based myoelectrically controlled biomimetic prosthetic/robotic hand (called Kobe hand) with three fingers—thumb, index, and middle fingers, was developed at Kobe University, Japan, by [53] with a system consisting of EMG signal processing unit, the dynamic model, positional control unit, and the prosthetic/robotic device. A survey of four important properties of dexterity, equilibrium, stability, and dynamic behavior relating to autonomous multi-fingered robotic hands was presented in [76]. An interesting aspect of this literature survey is a series of tables relating to existing multi-fingered robotic hands, force closure, dexterity in kinematically redundant robotic hands, equilibrium, in robotic grasp, and stability. As reported in [112], an intelligent prosthesis control system, developed by Animated Prosthetics, consists of two parts: the animation control system (ACS) residing in prosthesis and a remote prosthesis configuration unit (PCU) capable of on/off to variable speed/grip. Dynamic control of two arms to manipulate cooperatively an object with rolling contacts was addressed by [113] using a nonlinear feedback control methodology that decouples and linearizes the system.

A sensory control system based on force-sensing resistor (FSR) was developed by [114] at The National Institute for Accidents at Work (INAIL), Bologna, Italy, to control the strength of the grip on objects for a commercial prosthetic/robotic hand having two main functions: the automatic search for contact with the object and the detection of the object possibly slipping the grip by involuntary feedback (force sensors and slipping sensors). Further, automatic tuning of control parameters of prostheses was investigated by [115] using fuzzy logic (FL) expert systems resulting in a software package: microprocessor controlled arm auto tuning. The automatic tuning software works as follows: the client connects the prosthesis hardware, the program needs both sensor signals as client input, the program combines the above qualitative and quantitative information stored in the FL database to calculate the prosthesis parameter values, and the program enables the new parameter values to be down-loaded into the prosthesis control system memory. Dynamic modeling of a robotic hand was proposed in [116] using a hybrid approach with discrete event aspect of grasping and continuous-time part with a variable structure impedance control algorithm. A novel on-line learning method was reported by [56] for prosthetic/robotic hand control based on EMG measurements with a system consisting of three units: analysis unit for generating feature vectors containing useful information for discriminating motions from EMG signals, an adaptation unit for adapting to the amputee's individual variation and for discriminating motions from the feature vector and at the same time generating the necessary control commands to the prosthetic/robotic hand, and a trainer unit for directing the adaptation unit to learn in real time based on the amputee's teaching signal and the feature vector. The work by [114] built a sensory control system based on the FSR for an upper limb prosthesis and an optical sensor for detecting movement. The prostheses produced were of the "all or nothing" (opening or closing) and proportional control type (the relationship between force and EMG signal is linear). For traditional control, it used voluntary (visual) feedback, but the users had to pay good attention. This work developed an involuntary feedback control which uses two kinds of sensors, strength and slipping sensors. If the prosthesis hand is slipping, the control system automatically orders the actuator of the prosthesis to increase the grip strength. On receiving the EMG signal, the hand begins a closing action and goes on closing until the FSRs produce a signal that is greater than or equal to a "contact threshold" value, and then it stops, because the object has been grasped with the required strength of grip. The automatic grip mechanism is very useful in grasping delicate objects.

The investigation by [117] showed that the proposed neuro-fuzzy classifier known as Abe–Lan network, is able to identify correctly all the EMG signals related to different movements of human hand. A highly anthropomorphic human hand, called

Robonaut Hand consisting of five fingers and 14 independent DOFs, was built at NASA Johnson Space Center to interface with extra-vehicular activity (EVA) crew interfaces onboard International Space Station (ISS), as reported by [54].

2000–2007

In [118], estimating muscular contraction levels of flexors and extensors using neural networks (NNs), a new *impedance control* technique [119] was developed to control impedance parameters such as the moment of inertia, joint stiffness, and viscosity of a skeletal muscle model of a prosthetic/robotic hand. An overview of dextrous manipulation was provided by [78] with an interesting time-line chart for the development of robotic dextrous manipulation during the period 1960–2000. An excellent survey appears in [77] summarizing the evolution and state of the art in the robotic hands focusing mainly on functional requirements of manipulative dexterity, grasp robustness, and human operability. Also, the work by [120] exploited the non-holonomic character of a pair of bodies with regular rigid surfaces rolling onto each other, to study the constructive controllability algorithm for planning rolling motions for dextrous robot hands. A control system architectures was proposed in [121, 122] with a feedforward loop based on EMG measurements consisting of a low-pass filter and NN to provide the actual torque signal and a feedback loop based on desired angle consisting of a proportional-derivative (PD) controller to provide the desired torque signal and the error signal between these torques drives the prosthetic/robotic hand to achieve the desired angle while the NN learns based on feedback error.

This work reported by [123] studied finger extension, external control, overhead reach, and forearm pronation. For finger extension, they used two electrodes: one placed between the second and third metacarpals and the other between third and fourth metacarpals. They could provide full extension of the index, long, and ring fingers. For external control, a new form of control was developed by using retained voluntary wrist extension to control grasp opening and closing. Overhead reach is provided by stimulation of the triceps muscle, so elbow position is controlled by voluntary activation of biceps as an antagonist. As for forearm pronation, the main issues are an increased number of stimulus channels to allow stimulation of the finger intrinsic muscles, triceps, and forearm pronator, an implanted control source, bidirectional communication between sensor and body, reduced size, and reduction of all external cables. The work by [2] presents a review of the traditional methods for control of artificial hands using EMG signal, in both clinical and research areas and points out future developments in the control strategy of the prosthetics, in particular advocating neuroprosthesis with biocompatible neural interface for providing sensory feedback to the user leading to electroneurographic (ENG)-based control in place of EMG control. Collaboration between University of Southampton and University of New Brunswick (UNB) by [34] resulted in a hybrid control system using a multilayer perceptron (MLP) ANN as a classifier of time-domain features set (zero crossings, mean absolute value, mean absolute slope and trace length) extracted from MESs and a digital signal processor (DSP) controlling the grip pressure of the prosthetic/robotic hand without visual feedback (voluntary feedback). Design and development of an underactuated (the number of actuators less than the DOF) mechanism applicable to prosthetic/robotic hand was presented in [124] based on dynamic model of fingers leading to adaptive grasp (i.e., being able to conform to the shape of an object held within the hand).

Although an adaptive control scheme was developed by [94] for a Southampton Hand, further developments were made in the research by [79] and [125] producing their IP (Intelligent Prosthesis) according to [51]. The investigation [126] provided an evolution of microprocessor-based control systems for prosthetics with classification into first (based on digital systems), second (with low power), and third generation (based on microprocessors and DSPs). The work in [44] conducted a comparison of Oxford and Manus hand prostheses with respect to

1. hand mechanisms,

2. control electronics: EMG analog amplifiers, A/D converters, DSPs,

3. sensors: force, position and slip sensors based on Hall effect, and

4. manipulation or control schemes: Oxford hand used Southampton Adaptive Manipulation Scheme consisting of three-level hierarchical scheme and Manus used a two-level scheme.

The scheme suggested by [127] consisted of five modules, including an artificial muscoskeletal system, position and force sensors, 3D force sensors, low-level control loop dedicated to control slipping and grasping, and an EMG control unit. Further, the scheme used two semiconductor strain gauges as the force sensor and glues the sensor in SS496B by Honeywell International Inc. as the position sensor, which is the linear slider and small magnets. Moreover, the control system receives three signals: activation (EDG, which is used to identify whether there is a movement), direction (SGN, which decides opening or closing), and amplitude of the movement (AMP, which controls the seed of the movement in a proportional means). As for the control scheme, it uses a simple proportional open-loop control.

A cylindrical grasp of a cylindrical object and a parallel force/position control is studied by [128] to ensure the stability. The work in [129] presented a feedback control system for hand prosthesis with elbow control. Using a concept of extended physiological proprioception (EPP) (i.e., using natural physiological sensors), both the works [129] and the investigation by [130] developed microprocessor-based controllers for upper limb prostheses. A systematic literature review, conducted by [131], is useful for prosthetic/robotic hand, although the survey was done for lower limb prosthesis. This work by [128] developed a procedure to obtain maximum load and contact force distribution for a given grasp task and a parallel force/position control to ensure stability of the grasp. The goal of this control scheme is to specify a set of joint torque inputs so that the desired grasping forces along the constrained directions, and the desired position trajectory along the unconstrained directions are realized.

It was shown by [82, 132, 133] that sensory feedback signals are obtained for a multi-fingered robot hands to perform the function of grasping an object and that dynamic force/torque closure is constructed without knowing object kinematic parameters and location of the mass center. Further, the convergence of motion of the overall fingers-object system was proved using the concepts of "stability and asymptotic stability on a manifold." Mechanical design and manipulation (control) issues were addressed in [45] for a multi-fingered dextrous hand for upper limb prosthetics using the underactuated kinematics enhancing the performance and providing four grasping modes (cylindrical, precision, hook, and lateral) with just two actuators, one for the thumb and one for the remaining fingers. In particular, the hierarchical control architecture consists of a host (or master) controller for EMG management and definition of grasp set points (for position and torque/force) and three local (or slave) controllers for low level implementation of stiffness control of the joints. In [63], design and analysis was presented for a multi-fingered prosthetic/robotic hand consisting of a thumb with three joints and the rest of the four fingers having two joints using Haringx and element stiffness models, which enables the location of actuators far away from the hand to a belt around the waist and further enabling actuation and control with relatively high DOF. Robotic hand MA-I was designed and built by [64] at the Institute of Industrial and Control Engineering (IOC) at the Polytechnic University of Catalonia (UPC) with 16 degrees of freedom and the control system consisting of 16 position control loops, independently controlling each of the 16 DC motors. Visual hand motion capture is a multiple-dimension and multiple-objective searching optimization problem and the work reported in [134] used pose estimation and a motion-tracking scheme with genetic algorithms (GAs) embedded particle filter (PF) to navigate visual hand gesture, such as virtual environment and control of a robot arm.

The fabrication of a complaint, under-actuated prosthetic/robotic hand (both palm and fingers) moulded as a soft polymeric single part for providing *adaptive* grasp was reported by [135, 136]. Since the analysis and synthesis are "so complex and only experimental analysis of the solution adapted validate our works." It was shown by [137] that an object with parallel surfaces in a horizontal plane could be controlled by a pair of robotic fingers to achieve stable grasping, angle, and position control without the need for the object parameters or object sensors such as tactile, force, or visual sensors. At Northwester University Prosthetics Laboratory (NUPL), the researchers [138, 139] developed multi-function prosthetic/robotic hand/arm controller system receiving signals from as many as 16 implantable myoelectric sensors (IMES) and a heuristic FL approach to EMG signal pattern recognition by [140, 141]. In particular, FL was explored for discriminating between multiple surface EMG control signals and classify them to user intention. The multi-functional hand mechanism consisted of three motor hands (one motor for driving the thumb, one motor drives index finger, and the third motor drives middle, ring, and little finger) and two motor wrists (one motor for wrist extension/flexion and the other motor for wrist rotation). Further, the research by [67] demonstrated that in implementing the EPP control for a powered prosthesis, the backlash is determined by the stiffness of the control cable as well as mass located at the distal end of the forearm and that reduction of static friction and backlash in the system could prevent the limit cycle.

It was demonstrated by [142] that by implanting electrodes within individual fascicles of peripheral nerve stumps, appropriate, distally referred sensory feedback about joint position and grip force from an artificial arm could be provided to an amputee through stimulation of the severed peripheral nerves which also provide appropriate signals. It is interesting to note the work of [143] on the mechanism, design, and control system of a humanoid-type hand with human-like manipulation capabilities as a part of development of service robots and the comparison (shown by [144, 145]) of natural and prosthetic/robotic hands. In [146], the EMG motion pattern classifier was developed using on parametric autoregressive (AR) model and Levenberg–Marquardt (LM)-based NNs to identify three types of motion of thumb, index, and middle fingers to control a five-fingered underactuated prosthetic/robotic hand.

The work in [147] focussed on the "optimal" delay as the maximum amount of time, which is from command to hand movement, for a prosthesis controller with a delay of 200–400 ms as the range which is accepted by users. A bypass prosthesis called Prosthetic Hand for Able-Bodied Subjects (PHABS) was developed to allow able-bodied subjects to operate a prosthetic/robotic terminal device. The controller is a commercially available Myo-pulse control, which combines pulse width modulation (PWM) and pulse period modulation (PPM) because it provides a linear relation between motor speed and the pulse width and timing of a digital control signal. In addition, it also used a mechanical low-pass filter to smooth the pulse train and movement. If the EMG reaches the threshold, the motor will be turned "on"; otherwise, it will be turned "off." Furthermore, the experimental controller was created in Simulink of MATLAB and executed using Simulink Real Time and XPC Target Toolboxes. Finally, this work summarized seven time-delay sources, including

1. the time from the intent of movement to the development of EMG,

2. the time constant of the analog filters contained in the EMG pre-amplifiers,

3. the analog-to-digital sampling period,

4. the time required to collect the EMG signal for feature extraction,

5. the time required to perform the EMG signal for feature extraction,

6. the time required to execute th pattern recognition on the extracted features, and

7. the time required to actuate the component.

In [81], a review of the traditional methods of control as well as the current state of new control techniques was provided. A newly developed intelligent flexible hand system with 3 fingers, 10 joints, fitted with a small harmonic drive gear and a high power mini actuator, providing 12 DOFs applied to a catching task was developed by [148]. The authors [149] developed an EMG-based (using electrodes, torque, and angle sensors) prosthetic/robotic hand control system composed of a human operator, a five-fingered under-actuated prosthetic/robotic hand system, the prosthetic/robotic hand controller (with analog-to-digital converters and DSP board and stepper motors), and visual feedback. In particular, the EMG signals undergo feature extraction and feature classification using NNs with parametric autoregressive (AR) model and wavelet transforms. In an under-actuated system, there is less number of actuators compared to the number of DOF of the system. Further in [150], a hierarchical control system was proposed with a high-level supervisory controller for implementing the EMG signal acquisition and pattern recognition and also providing a set of commands (for operations such as close, open, position, etc.) to a low-level controller. A sensor-based hybrid control strategy (using normal feedback control based on EMG signals from sensors and feedback to the user) was presented by [151] where a digital controller operating from prosthetic signals converts the user grasping intention (EMG signal) into an order for the control of prosthesis.

The investigation by [68] developed a robot hand with tactile sensors (slip sensor and force sensor), called SKKU Hand II, having two functional units: a PolyVinyliDene Fluoride (PVDF)-based slip sensor designed to detect slippage and a thin flexible force sensor that read the contact force of and geometrical information on the object using a pressure variable resistor ink. A biomechatronic approach to the design and control of an anthropomorphic artificial hand was studied by [152] for closing the hand finger while grasping an object using a reference trajectory and using two different versions (joint space and slider space) of PD control system. In particular, the artificial hand consists of three under-actuated fingers (index, middle, and thumb) which are actuated by three cable-driven DC motors placed in the lower part of the arm. The work by [153] studied large controller delays created by multi-functional prosthesis controllers. A device called PHABS was utilized to test the performance of 20 able-bodied subjects to the Box and Block Test. To estimate and compare the performance of prosthetic/robotic hands, a functionality index is proposed by [147]. An underwater flexible robot manipulation (called HEU Hand II) that utilized Position-Based Neural Network Impedance Control (PBNNIC) for the force tracking control was studied by [154].

This work from [154] developed dextrous underwater robot hand, called as HEU Hand II. The sensor system mainly includes 12 strain gauges at different locations. When the robot hand is under water, the control system is more complicated because the complete dynamic model is not known exactly. Hence, the control system considers the uncertainty of the robot dynamic model. The controller of the hand force tracking is designed by PBNNIC scheme. Using biologically inspired principles for design and control of a bionic robot arm by [155], several control approaches were presented such as trajectory planning and optimization based on robot dynamics.

An alternate learning control strategy was proposed by [156] based on the working assumptions that both human motor commands and sensory information are passed on in a discrete, episodic manner, quantized in time with a learning algorithm called S-learning based on *sequences* arguing against the traditional control approaches due to highly nonlinear robot's dynamics and large number of DOF.

In the works by [157], the first prototype of a five-fingered prosthetic/robotic hand fitted with only three motors and achieving 20 DOFs was described using a new "strings and springs" mechanism and a continuous wavelet transform (CWT) for extraction of EMG inputs for a feed-forward, back-propagation NN to recognize the type of grip.

The work in [158] focuses on the control system of the hand and on the optimization of the hand design. It proposes the control action as proportional to the superficial EMG signals extracted by surface electrodes applied to a couple of antagonistic user's residual muscles. This work first explains designs of the hand prototype, such as biomechatronic design approach, under-actuated artificial hand, 3D CAD model (by ProEngineer), and dynamic analysis (by ANSYS). Secondly, it builds the model of control system, including the kinetics and dynamics of hand in PD control in the joint space and slider space with elastic compensation. Thirdly, it validates and optimizes the hand design in multiple objective problems (four goals). The first two goals are related the closed-loop control performance and the remaining two goals are part of joint trajectories. Besides, it develops the simulation in MATLAB/Simulink. Finally, it compares the experimental results with the simulation.

The dynamic system of a nonlinear flexible robot arm with a tip mass was introduced by [159] and the proposed intelligent optimal controller, in which the fuzzy neural network controller and robust controller were respectively designed to learn a nonlinear function and compensate the approximation errors, could control the coupling of bending vibration and torsional vibration for the periodic motion. To overcome the traditional FL difficulties, such as large rule bases and long training times, [160] proposed a self-learning dynamic fuzzy network (DFN) with dynamic equality constraints to speed up the trajectory calculations for intelligent nonlinear optimal control. For a five-finger under-actuated prosthetic/robotic hand with tendon transmission, [161] presented a robust controller implemented two subsequent and different phases, including the pre-shaping of the hand and the involved fingers rapidly closing around the object.

1.2.3   Overview of Main Control Techniques Since 2007

Hard Computing strategies:

1. **PD Controller**: Rong et al. [162] presented one kind of PD controller with feed-forward control based on adaptive theory for two DOFs direct driven robot with uncertain parameters.

2. **Adaptive Controller**: Cai et al. [163] developed an observer back-stepping adaptive control scheme for two-link manipulator under unmeasured velocity and uncertain environment and the adaptive velocity observer was designed independently from the state-feedback controller in order to compensate the estimation errors. Seo and Akella [164] derived the novel adaptive control solution involving a new filter design for the regressor matrix for n-DOF robot manipulator systems. By developing the Fourier series expansion from input reference signals of every joint, Liuzzo and Tomei [165] designed a global, output error feedback, adaptive learning control for two-DOF planar robot with uncertain dynamics. To achieve the tracking control objective, Chen et al. [166] proposed an adaptive sliding-mode dynamic controller for wheeled mobile robots with

system uncertainties and disturbances to make the real velocity of the wheeled mobile robot reach the desired velocity command.

3. **Robust Controller**: Because of the visco-elastic properties of manipulator links, Torabi and Jahed [167] utilized the loop-shaping method which decreases the order of the robust control model of a single-link manipulator examined in time and frequency domains. To enhance control of powered prosthetic/robotic hands, Engeberg and Meek [168–171] proposed robust sliding mode, back-stepping, and hybrid sliding mode-back-stepping (HSMBS) parallel force–velocity controllers which enabled the humans to more easily control a fine object by 10 able-bodied test subjects. Ziaei et al. [172] developed the modeling, system identification adopting generalized orthonormal basis functions (GOBFs), and robust position and force controllers for a single flexible link (SFL) manipulators required to operate the contact motion. Jiang and Ge [173] transformed the nonlinear kinematic models of three-DOF mobile robot with uncertain disturbance into linear control systems through an approximate linearization algorithm and then designed a partial feedback H_∞ robust controller through linear matrix inequality (LMI).

4. **Optimal Controller**: Vitiello et al. [174] synthesized the position controller and the Kalman filter to perform the planar movements, such as reaching and catching, of the NEURARM hydraulic piston actuation with nonlinear springs connected on the cable. Vrabie et al. [175] designed an online method via a biological inspired Actor/Critic structure to solve the adaptive optimal continuous-time control problem by the solution of the algebraic Riccati equation without using knowledge of the system internal dynamics. To minimize the positioning time (traveling between two specific points) of an under-actuated two-DOF robot manipulator restricted to the input constraint and the structural parameter constraint, Cruz-Villar et al. [176] developed a concurrent structure-control redesign method which combined the structural parameters and a bang–bang control law. Duchaine et al. [177] derived the position tracking and velocity control, the dynamic model of the robot, the prediction and control horizons, and the constraints by a general predictive control law and also derived an analytical solution for the optimal control by a computationally efficient model-based predictive control scheme for a six-DOF cable-driven parallel manipulator.

5. **Hierarchical Controller**: Fainekos et al. [178] proposed a hierarchical control law addressing the temporal logic motion planning problem for mobile robots modeled by second-order dynamics to track a simpler kinematic model with a globally bounded error and then the new robust temporal logic path planning problem for the kinematic model using automata theory and simple local vector fields were solved.

Soft Computing strategies:

1. **Fuzzy Logic**: According to human anatomy, Arslan et al. [179] developed the biomechanical model with a tendon configuration of the three-DOF index finger

of the human hand and the fuzzy sliding mode controller in which a FL unit tuned the slope of the sliding surface was introduced to generate the required tendon forces during closing and opening motion.

2. **Artificial Neural Networks**: Onozato and Maeda [180] utilized two NNs learning inverse kinematic and inverse dynamic to control the positions of two-DOF SCARA robot. Aggarwal et al. [181] obtained the neural recordings from rhesus monkeys with three different movements, the flexion/extension of each finger, the rotation of wrist and dextrous grasps and designed the separate decoding filters for each movement by using multilayer feed-forward ANN in order to be implemented in real-time MATLAB/Simulink. An online decentralized NN control design without deriving the dynamic model for a class of large-scale uncertain robot manipulator systems was proposed by Tan et al. [182]. Kato et al. [183] expressed the reaction of brains to the adaptable prosthetic/robotic system for a 13-DOF EMG signal controlled prosthetic/robotic hand with an EMG pattern recognition learning by ANNs. In addition, functional magnetic resonance imaging (f-MRI) was used to analyze the reciprocal adaptation between the human brain and the prosthetic/robotic hand by the plasticity of the motor and sensory cortex area in brains based on the variations in the phantom upper limb.

3. **Genetic Algorithm**: Marcos et al. [184] proposed the closed-loop pseudo-inverse method with genetic algorithms (CLGA) to minimize the largest joint displacement between two adjacent configurations, the total level of joint velocities, the joint accelerations, the total joint torque, and the total joint power consumption for the trajectory planning of three-DOF redundant robots. Kamikawa and Maeno [185] used GA to optimize locations of pivots and grasping force and designed one ultrasonic motor to move 15 compliant joints for an under-actuated five-finger prosthetic/robotic hand.

4. **Particle Swarm Optimization**: Khushaba et al. [186] developed a PSO-based method for myoelectrically controlled prosthetic/robotic devices. However, the artificial hands had limitation on precision grasping, such as grasping a screw or needle. To overcome the limitation, the accuracy and effectiveness of fingertip trajectory and control systems need to be optimized.

Fusion of Soft and Hard Computing strategies:

1. **PID Controller and Robust Controller**: Dieulot and Colas [187] presented a case study of the design of robust parametric methods for flexible axes and an heuristic initial tuning of the proportional-integral-derivative (PID) controller from additional pole placement constraints on the rigid mode.

2. **Adaptive Controller and Robust Controller**: To implement the trajectory tracking mission under the influence of unknown friction and uncertainty, Chen et al. [188] utilized a composite tracking scheme, including the adaptive friction

estimation to determine Coulomb friction, viscous friction, and the Stribeck effect and a robust controller to enhance the overall stability and robustness, for a two-DOF planar robot manipulator.

3. **Robust Controller and Optimal Controller**: Huang et al. [189] designed the robust control systems with some uncertainties, including the unknown payload and unknown modeling of objects and the unknown dynamic parameters, as the performance index that was optimized by the optimal control method for the space robot to capture unknown objects.

4. **Robust Controller and Fuzzy Logic**: Tootoonchi et al. [190] combined a robust quantitative feedback theory (QFT) designed to follow the desired trajectory tracking with the fuzzy logic controller (FLC) designed to reduce the complexities of the system dynamics for two-DOF arm manipulator. The control gain of the sliding mode controller tuned according to error states of the system by a fuzzy controller and a moving sliding surface whose the slope is dynamically changed by a FL algorithm for a three-DOF spatial robot were presented by Yagiz and Hacioglu [191].

5. **Robust Controller and Artificial Neural Networks**: Siqueira and Terra [192] developed a neural-network-based H_∞ controller which approximated the uncertain factors of an actual under-actuated cooperative manipulator and robustly controlled the position and squeeze force errors between the manipulator end-effectors and the object, although one joint was not actuated.

6. **Sliding Mode Controller and Genetic Algorithm**: Chen and Chang [193] utilized the multiple crossover GA to estimate the unknown system parameters and the sliding mode control method to overcome the uncertainty for a two-link robot control, respectively.

7. **Sliding Mode Controller and Particle Swarm Optimization**: Salehi et al. [194] used an online particle swarm optimization (PSO) to tune the parameters of sliding mode control at the contact moments of end-effector and unknown environments for the two-DOF planar manipulator.

8. **Fuzzy Logic and Artificial Neural Networks**: Subudhi and Morris [195] proposed a hybrid fuzzy neural control (HFNC) scheme containing a FLC and a NN controller to balance the coupling effects for the multi-link flexible manipulator with both rigid and flexible motions.

9. **Artificial Neural Networks and PSO**: Wen et al. [196] addressed the hybrid particle swarm optimization neural network (HPSONN) to compute the pseudo-inverse Jacobian of two-DOF planar manipulator inverse kinematic control.

1.2.4 Revolutionary Prosthesis

In 2009 (see the press releases [23, 69], the Applied Physics Laboratory (APL) of Johns Hopkins University (JHU), in Baltimore, MD received funding for the Rev-

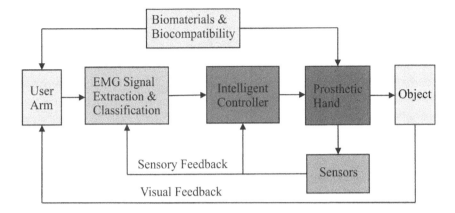

Figure 1.1   Schematic Diagram of Prosthetic/Robotic Hand Technology

olutionary Prosthesis 2009 program from DARPA (Defense Advanced Research Projects Agency), the U.S. Department of Defense, to "develop a next-generation mechanical arm that mimics the properties and sensory perception of the real thing." The APL leads an international team of about 30 organizations from Austria, Canada, Germany, Italy, Sweden, and USA. The APL team delivered first DARPA Limb Proto 1 (see [70], which "is a complete limb system that also includes a virtual environment used for patient training, clinical configuration, and to record limb movements and control signals during clinical investigations."

1.3   Fusion of Intelligent Control Strategies

Here we present the recent research activities on fusion control strategies for a smart prosthetic/robotic hand. The schematic diagram of the work is shown in Figure 1.1 (see the works of [34, 149, 151]). The overall system, in brief, consists of EMG signal acquisition from user arm for surface or implanted electrodes (in the implanted case we focus on biocompatibility based on nano-materials research). The EMG signal is then processed for feature extraction and classification or identification of EMG signal to correspond to different motions of the prosthetic/robotic hand. The classified signal is then used to control the prosthetic/robotic hand using actuators and driving mechanisms. It is to be noted that the EMG signal extraction and identification and the control algorithm are investigated using the fusion of soft computing (SC) and hard computing/control (HC) strategies.

1.3.1   Fusion of Hard and Soft Computing/Control Strategies

HC strategies are used at lower-level control for accuracy, precision, stability, and robustness and comprise PD control [197], PID control [198, 199], optimal control [199–202], adaptive control [203–206], etc. with specific applications to robotic hand devices. The authors conducted an overview of control strategies for robotic and prosthetic/robotic hands [85, 86]. However, our previous works [197–199, 207] for a robotic hand showed that PID controller resulted in undesirable feature of overshooting and oscillation, which were also demonstrated by Subudhi and Morris [195] in a two-link flexible robot manipulator and Liu and Chen [208] in a 6-DOF underwater robot (autonomous underwater vehicle).

The term SC or computational intelligence (CI) has been already used by L. A. Zadeh in 1994 and he defined SC as "a collection of methodologies that aim to exploit the tolerance for imprecision, uncertainty, partial truth, and approximation to achieve tractability, robustness, low solution cost and better rapport with reality" [209]. The fundamental concepts of SC have been influenced by Zadeh's earlier publications [210–212]. Since 1994, many researchers and engineers have worked on different methods using SC.

Unlike HC, SC strategies are meant to adapt to an environment under imprecision, uncertainty, partial truth, and approximation [209]. The review paper of L. Magdalena has analyzed, compared, and discussed some definitions of SC found in the literature [213]. Unlike the lower-level control of HC, SC is used at high-level control of the overall mission where human involvement and decision making is of primary importance. SC is an emerging field based on synergy and seamless integration of NN, FL, and optimization methods, such as GA and PSO [197, 209, 213–220]. The previous works on robotic/prosthetic hand used NN by [33, 34, 221], FL by [140, 141, 222], GA by [223], etc. mostly for EMG signal classification for various movements or functions of the robotic hand.

The brain analogy corresponds to the fusion of HC and SC strategies. We therefore propose hybrid intelligent control strategies with the integrated structure by blending [215, 216] the upper-level control of SC strategies and lower-level control of conventional HC strategies. Fusion of SC and HC methodologies can solve problems that cannot be solved satisfactorily by using either HC or SC methodology alone and can lead to high performance, robust, autonomous, and cost-effective missions, such as accuracy and effectiveness of fingertip trajectory and control systems [215, 216]. The hybrid intelligent control strategies for a robotic hand can be also applied to robotics for hazardous environments, surgery, etc. and clinical prosthetic/robotic hands [224–226].

The integration of SC and HC strategies shown in Figure 1.2 has the following attractive features [215, 216]:

1. The methodology based on SC is used, in particular with FL, at upper levels of the overall mission where human involvement and decision making is of primary importance, whereas the HC is used at lower levels for accuracy, precision, stability, and robustness.

2. In another situation using hybrid scheme, a NN of the SC is used to supplement the control provided by a linear, fixed gain controller for a missile autopilot.

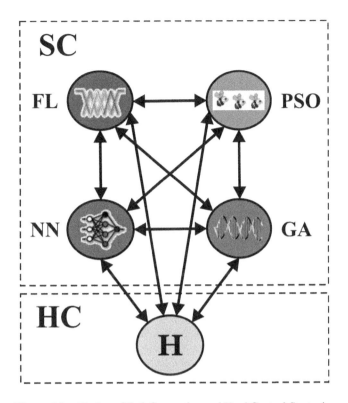

Figure 1.2   Fusion of Soft Computing and Hard Control Strategies

3. Further, the SC-based GA is used to tune the parameters of the PID controller and to achieve good performance and robustness for a wide range of operating conditions.

4. The SC and HC are potentially complementary methodologies.

5. The fusion could solve problems that cannot be solved satisfactorily by using either methodology alone.

6. Novel synergetic combinations of SC and HC lead to high performance, robust, autonomous, and cost-effective missions.

Our research focuses on developing intelligent autonomous strategies for EMG signal, extraction, analysis, and control of prosthetics by fusion of SC strategies comprising NN, FL, and GA (see [216, 227]) and HC strategies. The proposal takes advantage of our in-house research experience with problems in prosthetics as shown in [228, 229], in particular, and with problems in biomedical engineering as reported in [230, 231], in general.

An overview of nine papers using the strategies in industrial and engineering applications was presented by [232]. For the fusion strategies, the work by [233] described a multidimensional categorization scheme in five aspects: the degree of interconnection of soft and hard computing components (fusion grade), the topology of fusion skills (fusion structure), the time when fusion happens (fusion time), the layer of a system architecture (fusion level), and the motivation for the application (fusion incentive). Further, [234] classified the fusion strategies to 12 main categories and 6 supplementary categories.

1.4 Overview of Our Research

A chronological overview of our research is provided below.

A short review by Lai et al. [235] notes the importance of the biological interfaces that robotic implants and other prosthetic/robotic devices and notes an interdisciplinary team of biomedical and tissue engineers, and biomaterial and biomedical scientists is needed to work together holistically and synergistically.

In addressing the PSO technique, a set of operators for a PSO-based optimization algorithm is investigated for the purpose of finding optimal values for some of the classical benchmark problems. Particle swarm algorithms are implemented as mathematical operators inspired by the social behaviors of bird flocks and fish schools. In addition, particle swarm algorithms utilize a small number of relatively less complicated rules in response to complex behaviors, such that they are computationally inexpensive in terms of memory requirements and processing time. In particle swarm algorithms, particles in a continuous variable space are linked with neighbors, therefore the updated velocity of particles influences the simulation results. The work presents a statistical investigation on the velocity update rule for continuous variable PS algorithm. In particular, the probability density function influencing the particle velocity update is investigated along with the components used to construct the updated velocity vector of each particle within a flock. The simulation results of several numerical benchmark examples indicate that small amount of negative velocity is necessary to obtain good optimal values near global optimality [219].

A chronological overview of the applications of control theory to prosthetic/robotic hand is presented focusing on HC strategies such as multi-variable feedback, optimal, nonlinear, adaptive, and robust and SC strategies such as artificial intelligence, NN, FL, GA, PSO, and on the fusion of hard and soft control strategies [85]. The work [197] presents the PSO algorithm for identifying the rupture force for leukocyte adhesion molecules and the problem of finding the correct control parameters of a robotic hand. Another work by the group at ISU presents the fusion of SC technique of GA and HC technique of PID control with application to prosthetic/robotic hand. In particular, an adaptive neuro-fuzzy inference system (ANFIS) is used for inverse kinematics of the three-link index finger, and feedback linearization is used for the dynamics of the hand and the GA is used to find the optimal parameters of the PID controller [198]. An adaptive PSO (APSO) approach based on altering the maximum velocity at each iteration for two 30-dimensional benchmark problems is used [220].

A hybrid of a SC technique of ANFIS and a HC technique of adaptive control for a two-dimensional movement of a prosthetic/robotic hand with a thumb and index finger is investigated [205]. The dynamics of the prosthetic/robotic hand is derived and feedback linearization technique is used to obtain *linear* tracking error dynamics. Then the adaptive controller was designed to minimize the tracking error. The results of this hybrid controller show enhanced performance when compared with the PID controller. The adaptive control strategy is extended for the 14-DOF, five-fingered smart prosthetic/robotic hand with unknown mass and inertia of all the fingers [206]. The simulation results show that the five-fingered prosthetic/robotic hand with the proposed adaptive controller can grasp an object without overshooting and oscillation [236].

A novel condensed hybrid optimization (CHO) algorithm using enhanced continuous tabu search (ECTS) and the PSO was examined [207]. The proposed CHO algorithm combines the respective strengths of ECTS and the PSO. In particular, the ECTS is utilized to define smaller search spaces, which are used in a second stage by the basic PSO to find the respective local optimum. The ECTS covers the global search space by using a TS concept called diversification and then selects the most promising areas in the search space. Once the promising regions in the search space are defined, the proposed CHO algorithm employs another TS concept called intensification in order to search the promising area thoroughly. The proposed CHO algorithm is tested with the multi-dimensional hyperbolic and Rosenbrock problems. Compared to the other four algorithms, the results indicate that the accuracy and effectiveness of the proposed CHO algorithm was enhanced. Another hybrid of a SC technique using the ANFIS and a HC technique using finite-time linear quadratic optimal control for a two-fingered (thumb and index) prosthetic/robotic hand was investigated [201, 237, 238]. In particular, the ANFIS is used for inverse kinematics, and the optimal control is used to minimize tracking error utilizing feedback linearized dynamics. The simulations of this hybrid controller, when compared with the PID controller showed enhanced performance. This work was extended to a five-fingered, three-dimensional prosthetic/robotic hand [199]. To make the optimal controller fast acting and improve the accuracy, the performance index J is modified by including an exponential term [202, 239]. Simulations show that the proposed technique provides fast action with high accuracy and 30-fold faster than ANFIS- or GA-based trajectory planning [201, 239, 240].

1.5   Developments in Neuroprosthetics

It is worth noting some of the developments in neuroprosthesis reported in [241–244].

An interesting study was made by [245] on implanted neuroprostheses employing functional electric simulation (FES) to provide grasp and release to individuals with tetraplegia and comparing three control methods for shoulder position, wrist position, and myoelectric wrist extensors. To improve the control of grasp strength, forearm pronation, and elbow extension to the people with spinal cord injury at C5 and C6, the investigation by [123] developed an advanced neuroprosthesis that includes implanted components, including 10-channel stimulator, leads and electrodes, and a joint angle transducer, and external components, such as a control unit and transmitter–receiver coil.

In particular, it was reported in [246–248] that Jesse Sullivan, who lost both arms in an electric accident, could move his bionic arm with his brain—basically rewiring the severed live nerves that control arm and hand movements by redirecting the nerves to pectoral muscles in his chest. Electrodes attached to the chest muscles produce an electrical signal which controls the robotic arm depending upon the nature of muscle movement which in itself is characterized by "thinking" in the brain what is to be done with arms. However, the demonstrated bionic arm is only a "prototype and for research only."

Another interesting news appeared in [249, 250] regarding implantation of an electronic chip into the brain of a quadripledge man to use a computer to operate a robotic arm.

An article that appeared in IEEE Spectrum issue of September 2014 [251], describes about "an epilepsy patient ... controlling the mechanical limb with her brain waves."

1.6 Chapter Summary

This book is composed of seven chapters. Chapter 2 presents kinematics and trajectory planning and Chapter 3 presents dynamics for the robotic hand. The SC strategies such as FL, NN, ANFIS, GA, and PSO are addressed in Chapter 4. Chapters 5 and 6 present the fusion of soft and hard control strategies for each finger of the robotic hand and all the five fingers. Finally, conclusions and some thoughts on future work are given in Chapter 7.

Bibliography

[1] E. R. Kandel and J. H. Schartz. *Principles of Neural Science, Third Edition.* Elsevier/North-Holland, New York, USA, 1985.

[2] M. Zecca, S. Micera, M.C. Carrozza, and P. Dario. Control of multifunctional prosthetic hands by processing the electromyographic signal. *Critical Reviews^{TM} in Biomedical Engineering*, 30:459–485, 2002 (Review article with 96 references).

[3] MIT. *The Third Revolution: The Convergence of Life Sciences, Physical Sciences and Engineering.* Massachusetts Institute of Technology (MIT), Cambridge, Massachusetts, USA, 2011 (The First Revolution: Molecular and Cellular Biology and The Second Revolution: Genomics).

[4] College of Fellows American Institute for Medical and Biological Engineering. Medical and biological engineering in the next 20 years: The promise and

the challenges. *IEEE Transactions on Biomedical Engineering*, 60(7):1767–1775, July 2013.

[5] B. He, R. Baird, R. Butera, A. Datta, S. George, B. Hecht et al. Grand challenges in interfacing engineering with life sciences and medicine. *IEEE Transactions on Biomedical Engineering*, 60(3):589–598, March 2013.

[6] Committee on Key Challenge Areas for Convergence and Health. *Convergence: Facilitating Transdisciplinary Integration of Life Sciences, Physical Sciences, Engineering, and Beyond.* National Research Council (NRC) of the National Academies, Washington, District of Columbia, USA, 2014.

[7] What are the applications of an industrial robot arm? http://www.robots.com.

[8] M. Jamshidi and P. J. Eicker, editors. *Robotics and remote systems for hazardous environments.* Prentice Hall PTR, Upper Saddle River, New Jersey, USA, 1993.

[9] B. Brumson. Chemical and hazardous material handling robotics. Technical note, Robotic Industries Association, Ann Arbor, Michigan, USA, January 18, 2007 (http://www.robotics.org).

[10] S. Bouchard. Robotic arms help upgrade international space station. Technical note, IEEE Spectrum, New York, USA, July 22, 2009 (http://spectrum.ieee.org/automaton/robotics/humanoids).

[11] E. Guizzo. How robonaut 2 will help astronauts in space. Technical note, IEEE Spectrum, New York, USA, February 23, 2011 (http://spectrum.ieee.org/automaton/robotics/humanoids).

[12] Homeland Security Newswire, Mineola, New York, USA. *Homeland Security Newswire-Robotics*, 2015 (http://www.homelandsecuritynewswire.com/topics/robotics).

[13] U.S. Food and Drug Administration, Silver Spring, Maryland, USA. *Computer-Assisted Surgical Systems: Common uses of Robotically-Assisted Surgical (RAS) Devices*, March 11, 2015.

[14] Casualties in Afghanistan and Iraq. www.unknownnews.net, June 5, 2006.

[15] Amputee Coalition of America (ACA) National Limb Loss Information Center (NLLIC) Limb Loss Facts in the United States. http://www.amputee-coalition.org, March 10, 2011.

[16] P. F. Adams, G. E. Hendershot, and M. A. Marano. Current estimates from the national health interview survey, 1996 *National Center for Health Statistics. Vital and Health Statistics*, 200(10):1–203, 1999.

[17] K. Ziegler-Graham, E. J. MacKenzie, P. L. Ephraim, T. G. Travison, and R. Brookmeyer. Estimating the prevalence of limb loss in the united states:

2005 to 2050. *Archives of Physical Medicine and Rehabilitation*, 89:422–429, March 2008.

[18] D. Moniz. Arm Amputees Rely on Old Devices. USA Today:10-06-2005, June 2005.

[19] D. Moniz. Military to Fund Prosthetics Research. USA Today:10-06-2005, June 2005.

[20] Director of Defense Advanced Research Projects Agency (DARPA) Presentation to Subcommittee on Terrorism, Unconventional Threats and Capabilities, House Armed Services Committee, United States House of Representatives - Bridging the Gap. Press Release: Dated March 25, 2004.

[21] Director of Defense Advanced Research Projects Agency (DARPA): Bridging The Gap Powered by Ideas. Press Release: Dated February 2005.

[22] Director of Defense Advanced Research Projects Agency (DARPA): Defense Sciences Office, 2006.

[23] DARPA - News Release: DARPA Initiates Revolutionary Prosthetic Programs. Press Release: Dated February 8, 2006.

[24] E. Strickland. The end of disability: Prosthetics and neural interfaces will do away with biologys failings. *IEEE Spectrum*, 6:30–35, June 2014. (The Next Fifty Years: THE FUTURE WE DESERVE – A Special Report-BIOMED).

[25] IEEE, New York, USA. *IEEE Solutionists Showing What Engineers Do*, July 21, 2011. (https://ieeetv.ieee.org/ieeetv-specials/ieee-solutionists).

[26] M. Zecca, S. Roccella, G. Cappiello, K. Ito, K. Imanishi, H. Miwa, C. Carrozza, P. Dario, and A. Takanishi. From the human hand to a humanoid hand: Biologically-inspired approach for the development of robocasa hand #1. Technical report, 3ARTS Lab, Scuola Superiore Sant Anna, Pisa, Italy, 2006.

[27] A. E. Kobrinski. "Problems of bioelectric control," in *Proceedings of First IFAC*, p. 619, Moscow, USSR, 1960.

[28] E. D. Sherman. A Russian bioelectric-controlled prosthesis. *Canadian Medical Association Journal (CMAJ)*, 91:1268–1270, December 12, 1964.

[29] D. S. McKenzie. The Russian myo-electric arm. *The Journal of Bone and Joint Surgery*, 47:418–420, August 1965.

[30] I. Kato, E. Okazaki, H. Kikuchi, and K. Iwanami. "Electro-pneumatically controlled hand prosthesis using pattern recognition of myo-electric signals," in *Digest of 7th ICMBE*, p. 367, 1967.

[31] R. W. Mann and S. D. Reimers. Kinesthetic sensing for EMG controlled "Boston arm." *IEEE Transactions on Man-Machine Systems*, MMS-11:110–115, 1970.

[32] T. A. Rohland. Sensory feedback for powered limb prostheses. *Medical and Biological Engineering*, 12:300–301, March 1975.

[33] B. S. Hudgins. *A Novel Approach to Multifunctional Myoelectric Control of Prosthesis*. PhD Dissertation, University of New Brunswick, Fredericton, Canada, 1991.

[34] C. M. Light, P. H. Chappell, B. Hudgins, and K. Engelhart. Intelligent multifunction myoelectric control of hand prostheses. *Journal of Medical Engineering and Technology*, 26(4):139–146, July–August 2002.

[35] H. Hanafusa and H. Asada. "Stable prehension by a robot hand with elastic fingers," in M. Brady, J. M. Hollerbach, T. L. Johnson, T. Lozano-Pérez, and M. T. Mason, editors, *Robot Motion: Planning and Control*, pp. 322–335. MIT Press, Cambridge, Massachusetts, USA, 1977.

[36] E. F. R. Crossley and F. G. Umholtz. Design of a three-fingered hand. *Mechanism and Machine Theory*, 12:85–93, 1977.

[37] T. Okada. Computer control of multijointed finger system for precise object-handling. *IEEE Transaction on Systems, Man, and Cybernetics*, 12(3):289–299, 1982.

[38] S. C. Jacobsen, D. F. Knutti, R. T. Johnson, and H. H. Sears. Development of the Utah artificial arm. *IEEE Transactions on Biomedical Engineering*, BME-29(4):249–269, April 1982.

[39] S. C. Jacobsen, J. E. Wood, D. F. Knutti, K. B. Biggers, and E. K. Iversen. The version I Utah/MIT dextrous hand. In H. Hanafusa and H. Inoue, editors, *Robotics Research: The Second International Symposium*, pp. 301–308. MIT Press, Cambridge, Massachusetts, USA, 1985.

[40] E. Iversen, H. H. Sears, and S. C. Jacobsen. Artificial arms evolve from robots, or vice versa? *IEEE Control Systems Magazine*, 25(1):16–18, 20, February 2005.

[41] J. K. Salisbury. *Kinematic and Force Analysis of Articulated Hands*. PhD Dissertation, Stanford University, Stanford, California, USA, 1982.

[42] S. T. Venkataraman and T. E. Djaferis. "Multivariable feedback control of the JPL/stanford hand," in *Proceedings of the IEEE International Conference on Robotics and Automation*, pp. 77–82, 1987.

[43] D. Lian, S. Peterson, and M. Donath. "A three-fingered articulated hand," in *Proceedings of the 13th International Symposium on Industrial Robots*, pp. 18.91–18.101, 1983.

[44] P. J. Kyberd and J. L. Pons. "A comparison of the Oxford and Manus intelligent hand prostheses," in *Proceedings of the 2003 IEEE International Conference on Robotics and Automation*, pp. 3231–3236, Taipei, Taiwan, September 2003.

[45] J. L. Pons, E. Rocon, R. Ceres, D. Reynaerts, B. Saro, S. Levin, and W. Van Moorleghem. The MANUS-HAND dextrous robotics upper limb prosthesis: mechanical and manipulation aspects. *Autonomous Robots*, 16:143–163, 2004.

[46] H. Kobayashi. Control and geometric considerations for an articulated robot hand. *Journal of Robotic Research*, 1(1):3–12, 1985.

[47] A. Rovetta. "Sensors controlled multifingered robot hand," in *Proceedings of the IEEE Conference on Robotics and Automation*, pp. 1060–1063, St. Louis, Missouri, USA, March 1983.

[48] J. J. Kim, D. R. Blythe, D. A. Penny, and A. A. Goldenberg. "Computer architecture and low level control of the PUMA/RAL-hand system," in *Proceedings of the IEEE Conference on Robotics and Automation*, pp. 1590–1594, Raleigh, North Carolina, USA, March 1987.

[49] H. Van Brussel, B. Santoso, and D. Reynaerts. "Design and control of a multifingered hand provided with tactile feedback," in *Proceedings of the NASA Conference on Space Telerobotics*, pp. 89–101, Pasadena, California, USA, January/February 1989.

[50] G. A. Bekey, R. Tomovic, and I. Zeljkovic. "Control architecture for the belgrade/usc hand," in S. T. Venkataraman and T. Iberall, editors, *Dextrous Robot Hands*, pp. 136–149. Springer-Verlag, New York, USA, 1990.

[51] P. Kyberd and P. H. Chappell. The Southampton Hand: an intelligent myoelectric prosthesis. *Journal of Rehabilitation Research and Development*, 31(4):326–334, 1994.

[52] P. J. Kyberd, O. E. Holland, P. H. Chappell, S. Smith, R. Tregidgo, P. J. Bagwell, and M. Snaith. MARCUS: a two degree of freedom hand prosthesis with hierarchical grip control. *IEEE Transactions on Rehabilitation Engineering*, 3(1):70–76, March 1995.

[53] R. Okuno, M. Yoshida, and K. Akazawa. "Development of biomimetic prosthetic hand controlled by electromyogram," in *1996 4th International Workshop on Advanced Motion Control*, pp. 103–108, Mie, Japan, March 1996.

[54] C. S. Lovchik and M. A. Diftler. "The Robonaut hand: a dexterous robot hand for space," in *Proceedings of the IEEE International Conference on Robotics and Automation*, pp. 907–912, May 1999.

[55] H. Huang and C. Chen. "Development of a myoelectric discrimination system for a multi-degree prosthetic," in *Proceedings of the 1999 IEEE International Conference on Robotics and Automation*, pp. 2392–2397, Detroit, Michigan, USA, May 1999.

[56] D. Nishikawa, W. Yu, H. Yokoi, and Y. Kakazu. On-line learning method for EMG prosthetic hand control. *Electronics and Communications in Japan (Part III: Fundamental Electronic Science)*, 84(10):35–46, 2001 (Translated from Denshi Joho Tsushin Gakkai Ronbunshi, Vol. J82-D-II, No. 9, September 1999, pp.1510-1519).

[57] H. Liu, J. Butterfass, S. Knoch, P. Meusel, and G. Hirzinger. A new control strategy for DLR's multisensory articulated hand. *IEEE Control Systems Magazine*, 19(2):47–54, April 1999.

[58] J. Butterfass, M. Grebenstein, H. Liu, and G. Hirzinger. "DLR-Hand II: next generation of a dextrous robot hand," in *Proceedings of the IEEE International Conference on Robotics and Automation*, pp. 109–114, 2001.

[59] N. Fukaya, S. Toyama, T. Asfour, and R. Diffmann. "Design of the TUAT/Karlsruhe humanoid hand," in *Proceedings of the 2000 IEEE/RSJ International Conference on Intelligent Robots and Systems*, pp. 1–6, Stanford, California, USA, July 1996.

[60] Y. Zhang, Z. Han, H. Zhang, X. Shang, T. Wang, and W. Guo. "Design and control of the BUAA four-fingered hand," in *Proceedings of the 2001 IEEE International Conference on Robotics and Automation*, pp. 2517–2522, Seoul, South Korea, May 2001.

[61] N. Dechev, W. L. Cleghorn, and S. Naumann. Multiple finger, passive adaptive grasp prosthetic hand. *Mechanism and Machine Theory*, 36:1157–1173, 2001.

[62] R. Kolluru, K. P. Valavanis, P. Kimon, S. Smith, and N. Tsourveloudis. An overview of the University of Louisiana robotic gripper system project. *Transactions of the Institute of Measurement and Control*, 24(1):65–84, 2002.

[63] J. Yang, E. P. Pitarch, K.Abdel-Malek, A. Patrick, and L. Lindkvist. A multi-fingered hand prosthesis. *Mechanism and Machine Theory*, 39(6):555–581, June 2004.

[64] R. Suárez and P. Grosch. "Dexterous robotic hand MA-I software and hardware architecture," in *Proceedings of the Intelligent Manipulation and Grasping*, pp. 91–96, Genova, 2004.

[65] S. Roccella, M. C. Carrozza, G. Cappeiello, M. Zecca, H. Miwa, K. Itoh, and M. Matsumoto. "Design, fabrication and preliminary results of a novel anthropomorphic hand for humanoid robotics: RCH-1," in *Proceedings of the 2004 IEEE/RSJ International Conference on Intelligent Robots and Systems*, pp. 266–271, Sendal, Japan, September 28–October 2, 2004.

[66] F. Lotti, P. Tiezzi, G. Vassura, L. Biagiotti, G. Palli, and C. Melchiorri. "Development of UB hand 3: early results," in *Proceedings of the 2005 IEEE International Conference on Robotics and Automation*, pp. 4488–4493, Barcelona, Spain, April 2005.

[67] T. R. Farrell, R. F. Weir, C. W. Heckathorne, and D. S. Childress. The effect of static friction and backlash on extended physiological proprioception control of a powered prosthesis. *Journal of Rehabilitation Research and Development*, 42(3):327–342, May–June 2005.

[68] B. Choi, S. Lee, H. R. Choi, and S. Kang. "Development of anthropomorphic robot hand with tactile sensor: SKKU Hand II," in *Proceedings of the 2006 IEEE/RSJ International Conference on Intelligent Robots and Systems*, pp. 3779–3784, Beijing, P. R. China, October 9–15, 2006.

[69] APL to Lead Team Developing Revolutionary Prosthesis. Press Release: Dated February 9, 2006.

[70] Revolutionizing Prosthetics 2009 Team Delivers First DARPA Limb Prototype. Press Release: Dated April 26, 2007.

[71] D. J. Atkins, D. C. Y. Heard, and W. H. Donovan. Epidemiologic overview of individuals with upper limb loss and their reported research priorities. *Journal of Prosthetic and Orthotics*, 8(1):2–11, 1996.

[72] D. S. Childress. Closed-loop control in prosthetic systems: Historical perspective. *Journal Annals of Biomedical Engineering*, 8(4-6):293–303, July 1980 (45 references).

[73] R. N. Scott and P. A. Parker. Myoelectric prostheses: state of the art. *Journal of Medical Engineering and Technology*, 12(4):143–151, July/August 1988.

[74] R. A. Grupen, T. C. Henderson, and I. D. McMammon. A survey of general purpose manipulation. *International Journal of Robotics Research*, 8(1):38–62, 1989.

[75] H. H. Sears and J. Shaperman. Proportional myoelectric hand control: an evaluation. *American Journal of Physical Medicine and Rehabilitation*, 70(1):20–28, February 1991.

[76] K. B. Shimoga. Robot grasp synthesis algorithms: a survey. *The International Journal of Robotics Research*, 15:230–266, 1996 (Survey article with over 130 references).

[77] A. Bicchi. Hands for dexterous manipulation and robust grasping: a difficult road toward simplicity. *IEEE Transactions on Robotics and Automation*, 16(6):652–662, December 2000 (Summary article with 191 references).

[78] A. M. Okamura, N. Smaby, and M. R. Cutkosky. "An overview of dexterous manipulation," in *Proceedings of the IEEE International C2000 conference*

on Robotics and Automation, pp. 255–262, San Francisco, California, USA, April 2000 (52 references).

[79] P. Kyberd, P. H. Chappell, and D. Gow. Advances in the control of prosthetic arms: Guest Editorial. *Technology and Disability*, 15(2):57–61, 2003.

[80] A. Muzumdar, editor. *Powered Upper Limb Prostheses Control, Implementation and Clinical Application*. Springer-Verlag, New York, USA, 2004.

[81] D. P. J. Cotton, A. Cranny, P. H. Chappell, N. M. White, and S. P. Beeby. "Control strategies for a multiple degree of freedom prosthetic hand," in *Proceedings of The Institution of Measurement and Control UK ACC Control 2006 Symposium*, pp. 211–218, 2006.

[82] S. Arimoto. *Control Theory of Multi-fingered Hands: A Modeling and Analytical-Mechanics Approach for Dexterity and Intelligence*. Springer-Verlag, London, UK, 2008.

[83] L. Birglen, T. Laliberte, and C. Gosselin. *Underactuated Robotic Hands*. Springer Tracts in Advanced Robotics. Springer-Verlag, Berlin, Germany, 2008.

[84] T. Inoue and S. Hirai. *Mechanics and Control of Soft-fingered Manipulation*. Springer, New York, USA, 2009.

[85] D. S. Naidu, C.-H. Chen, A. Perez, and M. P. Schoen. "Control strategies for smart prosthetic hand technology: An overview," in *The 30th Annual International Conference of the IEEE Engineering Medicine and Biology Society (EMBS)*, pp. 4314–4317, Vancouver, Canada, August 20–24, 2008 (In *Top 20 Articles, in the Domain of Article 19163667, Since its Publication (2008)* according to *BioMedLib: "Who is Publishing in My Domain?,"* **Ranked as No. 8 of 20** as on August 1, 2014, **Ranked as No. 9 of 20** as on May 4, 2015 and as on June 2, 2015).

[86] D. S. Naidu and C.-H. Chen. "Automatic control techniques for smart prosthetic hand technology: An overview," in U. R. Acharya, F. Molinari, T. Tamura, D. S. Naidu, and J. Suri, editors, *Distributed Diagnosis and Home Healthcare (D₂H₂): Volume 2*, pp. 201–223. American Scientific Publishers, Stevenson Ranch, California, USA, 2011.

[87] S. Micera, J. Carpaneto, and S. Raspopovic. Control of hand prostheses using peripheral information. *IEEE Reviews in Biomedical Engineering*, 3:48–68, 2010 (Review article with 161 references).

[88] A. Cloutier and J. Yang. Design, control, and sensory feedback of externally powered hand prostheses: A literature review. *Critical Reviews in Biomedical Engineering*, 41(2):161–181, 2013.

[89] E. F. Murphy and A. B. Wilson. "Limb prosthetics and orthotics," in M. Clynes and J. H. Milsum, editors, *Biomedical Engineering Systems*, pp. 489–549. McGraw-Hill, New York, USA, 1970.

[90] N. Wiener. *Cybernetics or Control and Communication in the Animal and the Machine*. MIT Press, Cambridge, Massachusetts, USA, 1948. Second Edition, 1961; Paper Back Edition 1965.

[91] C. K. Battye, A. Nightingale, and J. Whillis. The use of myo-electric currents in the operation of prostheses. *Journal of Bone and Joint Surgery*, 37-B:506, 1955.

[92] A. H. Bottomley. Myo-electric control of powered prostheses. *The Journal of Bone and Joint Surgery*, 47-B(3):411–415, August 1965.

[93] A. H. Bottomley, G. Kingshill, P. Robert, D. Styles, P. H. Jilbert, J. W. Birtill, and J. R. Truscott. Prosthetic hand with improved control system for activation by electromyogram signals. Technical report, National Research Development Corporation, London, UK, December 1968. US Patent 3418662.

[94] R. W. Todd. *Adaptive Control of a Human Prosthesis*. PhD Dissertation, University of Southampton, Southampton, UK, 1969.

[95] G. N. Saridis and H. E. Stephanou. A hierarchical approach to the control of a prosthetic arm. *IEEE Transactions on Systems, Man and Cybernetics*, 7(6):407–420, June 1977.

[96] G. N. Saridis and T. P. Gootee. EMG pattern analysis and classification for a prosthetic arm. *IEEE Transactions on Biomedical Engineering*, BME-29(6):403–412, June 1982.

[97] S. Lee and G. N. Saridis. The control of a prosthetic arm by EMG pattern recognition. *IEEE Transactions on Automatic Control*, 29(4):290–302, April 1984.

[98] A. A. Cole, J. E. Hauser, and S. S. Sastry. Kinematic and control of multifingered hands with rolling contact. *IEEE Transactions on Automatic Control*, 34(4):398–404, April 1989.

[99] R. M. Murray, Z. Li, and S. S. Sastry. *A Mathematical Introduction to Robotic Manipulation*. CRC Press, Boca Raton, Florida, USA, 1994.

[100] C. J. Abu-Haj and N. Hogen. Functional assessment of control systems for cybernetic elbow prostheses-Part I: Description of the technique. *IEEE Transactions on Biomedical Engineering*, 37(11):1025–1036, November 1990.

[101] C. J. Abu-Haj and N. Hogen. Functional assessment of control systems for cybernetic elbow prostheses-Part II: Application of the technique. *IEEE Transactions on Biomedical Engineering*, 37(11):1037–1047, November 1990.

[102] G. P. Starr. Experiments in assembly using a dexterous hand. *IEEE Transactions on Robotics and Automation*, 6(3):342–347, June 1990.

[103] A. E. Hines, N. E. Owens, and P. E. Crago. Assessment of input-output properties and control of neuroprosthetic hand grasp. *IEEE Transactions on Biomedical Engineering*, 39(6):610–623, June 1992.

[104] B. S. Hudgins, P. Parker, and R. N. Scott. A new strategy for multifunction myoelectric control. *IEEE Transactions on Biomedical Engineering*, 40(1):82–94, January 1993.

[105] D. Graupe, J. Magnussen, and A. A. Beex. A microprocessor system for multifunctional control of upper-limb prostheses via myoelectric signal identification. *IEEE Transactions on Automatic Control*, 23(4):538–544, August 1978.

[106] D. Graupe, R. W. Liu, and G. S. Moschytz. "Applications of neural networks to medical signal processing," in *Proceedings of the 27th IEEE Conference on Decision and Control*, pp. 343–347, Austin, Texas, USA, December 1988.

[107] K. Englehart and B. Hudgins. A robust, real-time control scheme for multifunction myoelectric control. *IEEE Transactions on Biomedical Engineering*, 50(7):848–854, July 2003.

[108] K. Englehart, B. Hudgins, and A. D. C. Chan. Continuous multifunction myoelectric control using pattern recognition. *Technology and Disability*, 15(7):95–103, 2003.

[109] T. Iberall, G. Sukhatme, D. Beattie, and G. A. Bekey. "Control philosophy and simulation of a robotic hand as a model for prosthetic hands," in *Proceedings of the 1993 IEEE/RSJ International Conference on Intelligent Robots and Systems*, pp. 824–831, Yokohama, Japan, July 1993.

[110] T. Iberall, G. Sukhatme, D. Beattie, and G. A. Bekey. "Control philosophy for a simulated prosthetic hand," in *Proceedings of the Rehabilitation Engineering and Assistive Technology Society of North America (RESNA)*, pp. 12–17, Las Vegas, Nevada, USA, June 1993.

[111] T. Iberall, G. Sukhatme, D. Beattie, and G. A. Bekey. "On the development of EMG control for a prosthesis using a robotic hand," in *Proc. of the 1994 IEEE International Conference on Robotics and Automation*, pp. 1753–1758, San Diego, California, USA, May 1994.

[112] D. J. Bak. Control system matches prosthesis to patient. *Design News*, p. 68, June 1997.

[113] N. Sarkar, X. Yun, and V. Kumar. Dynamic control of 3-D rolling contacts in two-arm manipulation. *IEEE Transactions on Robotics and Automation*, 13(3):364–376, June 1977.

[114] A. Tura, C. Lamberti, A. Davalii, and R. Sacchetti. Experimental development of a sensory control system for an upper limb myoelectric prosthesis. *Journal of Rehabilitation Research and Development*, 35(1):14–26, January 1998.

[115] C. Bonivento and A. Davalli. Automatic tuning of myoelectric prosthesis. *Journal of Rehabilitation Research and Development*, 35(3):294–310, July 1998.

[116] T. Schlegl and M. Buss. "Hybrid closed-loop control of robotic hand regrasping," in *Proceedings of the 1998 IEEE International Conference on Robotics and Automation*, pp. 3026–3031, Leuven, Belgium, May 1998.

[117] S. Micera, A. M. Sabatini, P. Dario, and B. Rossi. A hybrid approach to EMG pattern analysis for classification of arm movements using statistical and fuzzy techniques. *Medical Engineering Physics*, 21(5):303–311, June 1999.

[118] T. Tsuji, O. Fukuda, H. Shigeyoshi, and M. Kaneko. "Bio-mimetic impedance control of and EMG-controlled prosthetic hand," in *Proceedings of the 2000 IEEE/RSJ International Conference on Intelligent Robots and Systems*, pp. 377–382, Takamatsu, Japan, 2000.

[119] N. Hogan. Impedance control: An approach to manipulation, Part I, Part II and Part III. *Transactions of ASME, Journal of Dynamic Systems, Measurement, and Control*, 107:1–24, March 1985.

[120] A. Marigo and A. Bicchi. Rolling bodies with regular surface: Controllability theory and applications. *IEEE Transactions on Automatic Control*, 45(9):1586–1599, September 2000.

[121] S. Morita, K. Shibata, X. Z. Zheng, and K. Ito. "Prosthetic hand control based on torque estimation from EMG signals," in *Proceedings of the 2000 IEEE/RSJ International Conference on Intelligent Robots and Systems*, pp. 389–394, Takamatsu, Japan, October 31 – November 5, 2000.

[122] S. Morita, T. Kondo, and K. Ito. "Estimation of forearm movement from EMG signal and application to prosthetic hand control," in *Proceedings of the 2001 ICRA/IEEE International Conference on Robotics and Automation*, pp. 1477–1482, Seoul, South Korea, May 21–25, 2001.

[123] P. H. Peckham, K. L. Kilgore, M. W. Keith, A. M. Bryden, N. Bhadra, and F. W. Montague. An advanced neuroprosthesis for restoration of hand and upper arm control using an implantable controller. *The Journal of Hand Surgery*, 27A(2):265–276, March 2002.

[124] B. Massa, S. Roccella, M. C. Carrozza, and P. Dario. "Design and development of an underactuated prosthetic hand," in *Proceedings of the 2002 IEEE International Conference on Robotics and Automation*, pp. 3374–3379, Washington, District of Columbia, May 2002.

[125] Blatchford intelligent prosthesis. http://www.blatchford.co.uk, 1994.

[126] C. Lake and J. M. Miguelez. Evolution of microprocessor based control systems in upper extremity prosthetics. *Technology and Disability*, 15(2):63–71, 2003.

[127] M. C. Carrozza, F. Vecchi, F. Sebastiani, G. Cappiello, S. Roccella, M. Zecca, R. Lazzarini, and P. Dario. "Experimental analysis of an innovative prosthetic hand with proprioceptive sensors," in *Proceedings of the 2003 IEEE International Conference on Robotics and Automation*, pp. 2230–2235, September 14–19, 2003.

[128] P. Scherillo, B. Siciliano, L. Zollo, M. C. Carrozza, E. Guglielmelli, and P. Dario. "Parallel force/position control of a novel biomechatronic hand prosthesis," in *Proceedings of the 2003 IEEE/ASME International Conference on Advanced Intelligent Mechatronics (AIM 2003)*, pp. 920–925, 2003.

[129] D. H. Plettenburg and J. L. Herder. Voluntary closing: a promising opening in hand prosthetics. *Technology and Disability*, 15(2):85–94, 2003.

[130] H. M. Al-angari, R. F. ff. Weir, C. W. Heckanthorne, and D. S. Childress. A two-degree-of-freedom microprocessor based extended physiological proprioception (EPP) controller for upper limb prostheses. *Technology and Disability*, 15(2):113–127, 2003.

[131] H. van der Linde, C. J. Hofstad, A. C. H. Geurts, K. Postema, J. H. B. Geertzen, and J. van Limbeek. A systematic literature review of the effect of different prosthetic components of human functioning with a lower-limb prosthesis. *Journal of Rehabilitation Research and Development*, 41(4):555–570, July–August 2004 (Review article with 91 references).

[132] S. Arimoto, P. T. A. Nguyen, H.-Y. Han, and Z. Doulgeri. Dynamics and control of a set of dual fingers with soft tips. *Robotica*, 18:71–80, 2000.

[133] S. Arimoto. Intelligent control of multi-fingered hands. *Annual Reviews in Control*, 28:75–85, 2004.

[134] J. Cui and Z. Sun. "Visual hand motion capture for guilding a dexterous hand," in *Proceedings of the Sixth IEEE International Conference on Automatic Face and Gesture Recognition (FGR'04)*, pp. 729–734, May 17–19, 2004.

[135] M. C. Carrozza, B. Massa, S. Micera, R. Lazzarini, and M. Zecca. The development of a novel prosthetic handongoing research and preliminary results. *IEEE/ASME Transactions on Mechatronics*, 7(2):108–114, 2002.

[136] M. C. Carrozza, G. Cappiello, G. Stellin, F. Zaccone, F. Vechhi, S. Micera, and P. Dario. "On the development of a novel adaptive prosthetic hand with complaint joints: experimental platform and EMG control," in *Proceedings*

of the IEEE/RSJ International Conference on Intelligent Robots and Systems, pp. 1271–1276, August 2005.

[137] R. Ozawa, S. Arimoto, S. Nakamura, and J.-H. Bae. Control of an object with parallel surfaces by a pair of finger robots without object sensing. *IEEE Transactions on Robotics,* 21(5):965–976, October 2005.

[138] R. F. Weir. *Direct Muscle Attachment as a Control Input for a Position-Servo Prosthesis Controller.* PhD Dissertation, Northwestern University, Evanston, Illinois, USA, 1995.

[139] R. F. Weir, P. R. Troyk, G. DeMichele, and D. Kerns. "Technical details of the implantable myoelectric sensor (IMES) system for multifunction prosthesis control," in *Proceedings of the 25th IEEE Engineering in Medicine and Biology 27th Annual Conference,* pp. 7337–7340, Shanghai, P. R. China, September 2005.

[140] R. F. Weir and A. B. Ajiboye. "A multifunction prosthesis controller based on fuzzy-logic techniques," in *Proceedings of the 25th Annual International Conference of IEEE EMBS,* pp. 17–21, Cancun, Mexico, September 2003.

[141] A. B. Ajiboye and R. F. Weir. A heuristic fuzzy logic approach to EMG pattern recognition for multifunctional prosthesis control. *IEEE Transactions on Neural Systems and Rehabilitation Engineering,* 13(3):280–291, September 2005.

[142] G. S. Dhillon and K. W. Horch. Direct neural sensory feedback and control of a prosthetic arm. *IEEE Transactions on Neural Systems and Rehabilitation Engineering,* 13(4):468–472, December 2005.

[143] A. Kargov, T. Asfour, C. Pylatiuk, R. Oberle, H. Klosek, S. Schulz, K. Regenstein, G. Bretthauer, and R. Dillmann. "Development of an anthropomorphic hand for a mobile assistive robot," in *9th International Conference on Rehabilitation Robotics (ICORR),* pp. 182–186, Chicago, Illinois, USA, June–July 2005.

[144] A. Kargov, C. Pylatiuk, J. Martin, S. Schulz, and L. Derlein. A comparison of the grip force distribution in natural hands and in prosthetic hands. *Disability And Rehabilitation,* 26(12):705–711, 2004.

[145] C. Pylatiuk, S. Mounier, A. Kargov, S. Schulz, and G. Bretthauer. "Progress in the development of a multifunctional hand prosthesis," in *Proceedings of the 26th Annual International Conference of the IEEE EMBS,* pp. 4260–4263, San Francisco, California, USA, September 1–5, 2004.

[146] J. Zhao, Z. Xie, L. Jiang, H. G. Cai, H. Liu, and G. Hirzinger. "Levenberg-Marquardt based neural network control for a five-fingered prosthetic hand," in *Proceedings of the 2005 IEEE International Conference on Robotics and Automation,* pp. 1–6, Barcelona, Spain, April 2005.

[147] L. E. Rodriguez-Cheu, A. Casals, A. Cuxart, and A. Parra. "Towards the definition of a functionality index for the quantitative evaluation of hand-prosthesis," in *2005 IEEE/RSJ International Conference on Intelligent Robots and Systems (IROS)*, pp. 541–546, August 2–6, 2005.

[148] X.-T. Le, W.-G Kim, B.-C. Kim, S.-H. Han, J.-G. Ann, and Y.-H. Ha. "Design of a flexible multifingered robotic hand with a 12 D.O.F. and its control applications," in *Proceedings of the SICE-ICASE International Joint Conference*, pp. 1–5, Bexco, Busan, South Korea, October 18–21, 2006.

[149] J. Zhao, Z. Xie, L. Jiang, H. G. Cai, H. Liu, and G. Hirzinger. "A five-fingered underactuated prosthetic hand control scheme," in *Proceedings of the First IEEE/RAS-EMBS 2006 International Conference on Biomedical Robotics and Biomechatronics*, pp. 995–1000, Pisa, Italy, February 20–22, 2006.

[150] D. W. Zhao, L. Jiang, H. Huang, M. H. Jin, H. G. Cai, and H. Liu. "Development of a multi-DOF anthropomorphic prosthetic hand," in *Proceedings of the 2006 IEEE International Conference on Robotics and Biomimetics*, pp. 1–6, Kunming, P. R. China, December 17–20, 2006.

[151] L. E. Rodriguez-Cheu and A. Casals. "Sensing and control of a prosthetic hand with myoelectric feedback," in *Proceedings of the First IEEE/RAS-EMBS 2006 International Conference on Biomedical Robotics and Biomechatronics*, pp. 607–612, Pisa, Italy, February 20–22, 2006.

[152] L. Zollo, S. Roccella, R. Tucci, B. Siciliano, E. Guglielmelli, M. C. Carrozza, and P. Dario. "Biomechatronic design and control of an anthropomorphic artificial hand for prosthetics and robotic applications," in *Proceedings of the First IEEE/RAS-EMBS 2006 International Conference on Biomedical Robotics and Biomechatronics*, pp. 402–407, Pisa, Italy, February 20–22, 2006.

[153] T. R. Farrell and R. F. Weir. The optimal controller delay for myoelectric prostheses. *IEEE Transactions on Neural Systems and Rehabilitation Engineering*, 15(1):111–118, March 2007.

[154] Q. Meng, H. Wang, P. Li, L. Wang, and Z. He. "Dexterous underwater robot hand: HEU Hand II," in *Proceedings of the 2006 IEEE International Conference on Mechatronics and Automation*, pp. 1477–1482, Luoyang, P. R. China, June 25–28, 2006.

[155] S. Klug, O. von Stryk, and B. Mohl. "Design and control mechanisms for a 3 DOF bionic manipulator," in *Proceedings of the First IEEE/RAS-EMBS 2006 International Conference on Biomedical Robotics and Biomechatronics*, pp. 450–454, Pisa, Italy, February 20–22, 2006.

[156] B. Rohrer and S. Hulet. "A learning and control approach based on the human neuromotor system," in *Proceedings of the First IEEE/RAS-EMBS 2006*

International Conference on Biomedical Robotics and Biomechatronics, pp. 57–61, Pisa, Italy, February 20–22, 2006.

[157] J. Žajdlik. "The preliminary design and motion control of a five-fingered prosthetic hand," in *Proceedings of the 2006 International Conference on Intelligent Engineering Systems*, pp. 202–206, 2006.

[158] L. Zollo, S. Roccella, E. Guglielmelli, M. C. Carrozza, and P. Dario. Biomechatronic design and control of an anthropomorphic artificial hand for prosthetic and robotic applications. *IEEE/ASME Transactions on Mechatronics*, 12(4):418–429, August 2007.

[159] R.-J. Wai and M.-C. Lee. Intelligent optimal control of single-link flexible robot arm. *IEEE Transactions on Industrial Electronics*, 51(1):201–220, 2004.

[160] Y. Becerikli, Y. Oysal, and A. F. Konar. Trajectory priming with dynamic fuzzy networks in nonlinear optimal control. *IEEE Transactions on Neural Networks*, 15(2):383–394, 2004.

[161] C. Cipriani, F. Zaccone, G. Stellin, L. Beccai, G. Cappiello, M. C. Carrozza, and P. Dario. "Closed-loop controller for a bio-inspired multi-fingered underactuated prosthesis," in *Proceedings 2006 IEEE International Conference on Robotics and Automation (ICRA 2006)*, pp. 2111–2116, Orlando, Florida, USA, May 15–19, 2006.

[162] P. X. Rong, Z. J. He, C. D. Zong, and N. Liu. "Trajectory tracking of robot based on adaptive theory," in *2008 International Conference on Intelligent Computation Technology and Automation (ICICTA 2008)*, pp. 298–301, Changhsa, Hunan, P. R. China, October 20–22, 2008.

[163] J. Cai, X. Ruan, and X. Li. "Output feedback adaptive control of uncertainty robot using observer backstepping," in *2008 International Conference on Intelligent Computation Technology and Automation (ICICTA 2008)*, pp. 404–408, Changhsa, Hunan, P. R. China, October 20–22, 2008.

[164] D. Seo and M. R. Akella. Non-certainty equivalent adaptive control for robot manipulator systems. *Systems and Control Letters*, 58(4):304–308, April 2009.

[165] S. Liuzzo and P. Tomei. Prosthetic hand finger control using fuzzy sliding modes. *International Journal of Adaptive Control and Signal Processing*, 23:97–109, 2009.

[166] C.-Y. Chen, T.-H. S. Li, Y.-C. Yeh, and C.-C. Chang. Design and implementation of an adaptive sliding-mode dynamic controller for wheeled mobile robots. *Mechatronics*, 19:156–166, 2009.

[167] M. Torabi and M. Jahed. "A novel approach for robust control of single-link manipulators with visco-elastic behavior," in *Tenth International Conference on Computer Modeling and Simulation (UKSIM 2008)*, pp. 685–690, Cambridge, UK, April 1–3, 2008.

[168] E. D. Engeberg and S. G. Meek. "Adaptive object slip prevention for prosthetic hands through proportional-derivative shear force feedback," in *Proceedings of the 2008 IEEE/RSJ International Conference on Intelligent Robots and Systems*, pp. 1940–1945, Nice, France, September 22–26, 2008.

[169] E. D. Engeberg and S. G. Meek. Improved grasp force sensitivity for prosthetic hands through force-derivative feedback. *IEEE Transactions on Biomedical Engineering*, 55(2):817–821, 2008.

[170] E. D. Engeberg and S. G. Meek. Backstepping and sliding mode control hybridized for a prosthetic hand. *IEEE Transactions on Neural Systems and Rehabilitation Engineering*, 17(1):70–79, 2009.

[171] E. D. Engeberg, S. G. Meek, and M. A. Minor. Hybrid force velocity sliding mode control of a prosthetic hand. *IEEE Transactions on Biomedical Engineering*, 55(5):1572–1581, 2008.

[172] K. Ziaei, L. Ni, and D. W. L. Wang. Qft-based design of force and contact transition controllers for a flexible link manipulator. *Control Engineering Practice*, 17:329–344, 2009.

[173] W. Jiang and W. Ge. "Modeling and H_∞ robust control for mobile robot," in *2008 IEEE Conference on Robotics, Automation and Mechatronics (RAM 2008)*, pp. 1108–1112, Chengdu, P. R. China, September 21–24, 2008.

[174] N. Vitiello, E. Cattin, S. Roccella, F. Giovacchini, F. Vecchi, M. C. Carrozza, and P. Dario. "The neurarm: towards a platform for joint neuroscience experiments on human motion control theories," in *Proceedings of the 2007 IEEE/RSJ International Conference on Intelligent Robots and Systems*, pp. 1852–1857, San Diego, California, USA, October 29–November 2, 2007.

[175] D. Vrabie, F. Lewis, and M. Abu-Khalaf. Biologically inspired scheme for continuous-time approximate dynamic programming. *Transactions of the Institute of Measurement and Control*, 30:207–223, 2008.

[176] C. A. Cruz-Villar, J. Alvarez-Gallegos, and M. G. Villarreal-Cervantes. Concurrent redesign of an underactuated robot manipulator. *Mechatronics*, 19:178–183, 2009.

[177] V. Duchaine, S. Bouchard, and C. M. Gosselin. Computationally efficient predictive robot control. *IEEE/ASME Transactions on Mechatronics*, 12(5):570–578, 2007.

[178] G. E. Fainekos, A. Girard, H. Kress-Gazit, and G. J. Pappas. Temporal logic motion planning for dynamic robots. *Automatica*, 45:343–352, 2009.

[179] Y. Z. Arslan, Y. Hacioglu, and N. Yagiz. Prosthetic hand finger control using fuzzy sliding modes. *Journal of Intelligent and Robotic Systems*, 52:121–138, 2008.

[180] K. Onozato and Y. Maeda. "Learning of inverse-dynamics and inverse-kinematics for two-link scara robot using neural networks," in *The Society of Instrument and Control Engineers (SICE) Annual Conference 2007*, pp. 1031–1034, Kagawa University, Takamatsu, Japan, September 17–20, 2007.

[181] V. Aggarwal, G. Singhal, J. He, M. H. Schieber, and N. V. Thakor. "Towards closed-loop decoding of dexterous hand movements using a virtual integration environment," in *30th Annual International IEEE Engineering in Medicine and Biology Society Conference (EMBC 2008)*, pp. 1703–1706, Vancouver, British Columbia, Canada, August 20–24, 2008.

[182] K. K. Tan, S. Huang, and T. H. Lee. Decentralized adaptive controller design of large-scale uncertain robotic systems. *Automatica*, 45:161–166, 2009.

[183] R. Kato, H. Yokoi, A. H. Arieta, W. Yub, and T. Arai. Mutual adaptation among man and machine by using f-MRI analysis. *Robotics and Autonomous Systems*, 57:161–166, 2009.

[184] M. da G. Marcos, J. A. T. Machado, and T.-P. Azevedo-Perdicoulis. Trajectory planning of redundant manipulators using genetic algorithms. *Communications in Nonlinear Science and Numerical Simulation*, 14(7):2858–2869, July 2009.

[185] Y. Kamikawa and T. Maeno. "Underactuated five-finger prosthetic hand inspired by grasping force distribution of humans," in *Proceedings of the 2008 IEEE/RSJ International Conference on Intelligent Robots and Systems*, pp. 717–722, Nice, France, September 22–26, 2008.

[186] R. N. Khushaba, A. Al-Ani, and A. Al-Jumaily. "Swarm intelligence based dimensionality reduction for myoelectric control," in *Proceedings of the IEEE Conference on Intelligent Sensors, Sensor Networks and Information Processing*, pp. 577–582, Melbourne, Australia, December 3–6, 2007.

[187] J.-Y. Dieulot and F. Colas. Robust pid control of a linear mechanical axis: A case study. *Mechatronics*, 19:269–273, 2009.

[188] C.-Y. Chen, M. H.-M. Cheng, C.-F. Yang, and J.-S. Chen. "Robust adaptive control for robot manipulators with friction," in *The Third International Conference on Innovative Computing Information and Control (ICICIC '08)*, pp. 422–426, Dalian, Liaoning, P. R. China, June 18–20, 2008.

[189] P. Huang, J. Yan, J. Yuan, and Y. Xu. "Robust control of space robot for capturing objects using optimal control method," in *Proceedings of the 2007 International Conference on Information Acquisition (ICIA '07)*, pp. 397–402, Jeju City, South Korea, July 8–11, 2007.

[190] A. A. Tootoonchi, M. R. Gharib, and Y. Farzaneh. "A new approach to control of robot," in *2008 IEEE Conference on Robotics, Automation and Mechatronics (RAM 2008)*, pp. 649–654, Chengdu, P. R. China, September 21–24, 2008.

[191] N. Yagiz and Y. Hacioglu. Robust control of a spatial robot using fuzzy sliding modes. *Mathematical and Computer Modelling*, 49:114–127, 2009.

[192] A. A. G. Siqueira and M. H. Terra. Neural network-based H_∞ control for fully actuated and underactuated cooperative manipulators. *Control Engineering Practice*, 17:418–425, 2009.

[193] J. L. Chen and W.-D. Chang. Feedback linearization control of a two-link robot using a multi-crossover genetic algorithm. *Expert Systems with Applications*, 36:4154–4159, 2009.

[194] M. Salehi, G.R. Vossoughi, M. Vajedi, and M. Brooshaki. "Impedance control and gain tuning of flexible base moving manipulators using PSO method," in *Proceedings of the 2008 IEEE International Conference on Information and Automation*, pp. 458–463, Zhangjiajie, P. R. China, June 20–23, 2008.

[195] B. Subudhi and A. S. Morris. Soft computing methods applied to the control of a flexible robot manipulator. *Applied Soft Computing*, 9:149–158, 2009.

[196] X. Wen, D. Sheng, and J. Huang. *A Hybrid Particle Swarm Optimization for Manipulator Inverse Kinematics Control*, volume 5226 of *Lecture Notes in Computer Science*. Springer-Verlag, Berlin, Germany, 2008.

[197] C.-H. Chen, K. W. Bosworth, M. P. Schoen, S. E. Bearden, D. S. Naidu, and A. Perez-Gracia. "A study of particle swarm optimization on leukocyte adhesion molecules and control strategies for smart prosthetic hand," in *2008 IEEE Swarm Intelligence Symposium (IEEE SIS08)*, St. Louis, Missouri, USA, September 21–23, 2008.

[198] C.-H. Chen, D. S. Naidu, A. Perez-Gracia, and M. P. Schoen. "Fusion of hard and soft control techniques for prosthetic hand," in *Proceedings of the International Association of Science and Technology for Development (IASTED) International Conference on Intelligent Systems and Control (ISC 2008)*, pp. 120–125, Orlando, Florida, USA, November 16–18, 2008.

[199] C.-H. Chen, D. S. Naidu, A. Perez-Gracia, and M. P. Schoen. "A hybrid control strategy for five-fingered smart prosthetic hand," in *Joint 48th IEEE Conference on Decision and Control (CDC) and 28th Chinese Control Conference (CCC)*, pp. 5102–5107, Shanghai, P. R. China, December 16–18, 2009.

[200] D. S. Naidu. *Optimal Control Systems*. CRC Press, a Division of Taylor & Francis, Boca Raton, Florida, USA and London, UK, 2003 (A vastly expanded and updated version of this book, is under preparation for publication in 2017).

[201] C.-H. Chen, D. S. Naidu, A. Perez-Gracia, and M. P. Schoen. "A hybrid optimal control strategy for a smart prosthetic hand," in *Proceedings of the ASME 2009 Dynamic Systems and Control Conference (DSCC)*, Hollywood, California, USA, October 12–14, 2009 (No. DSCC2009–2507).

[202] C.-H. Chen and D. S. Naidu. "Optimal control strategy for two-fingered smart prosthetic hand," in *Proceedings of the International Association of Science and Technology for Development (IASTED) International Conference on Robotics and Applications (RA 2010)*, pp. 190–196, Cambridge, Massachusetts, USA, November 1–3, 2010.

[203] F. L. Lewis, S. Jagannathan, and A. Yesildirek. *Neural Network Control of Robotic Manipulators and Nonlinear Systems*. Taylor & Francis, London, UK, 1999.

[204] F. L. Lewis, D. M. Dawson, and C. T. Abdallah. *Robot Manipulators Control: Second Edition, Revised and Expanded*. Marcel Dekker, Inc., New York, USA, 2004.

[205] C.-H. Chen, D. S. Naidu, A. Perez-Gracia, and M. P. Schoen. "A hybrid adaptive control strategy for a smart prosthetic hand," in *The 31st Annual International Conference of the IEEE Engineering Medicine and Biology Society (EMBS)*, pp. 5056–5059, Minneapolis, Minnesota, USA, September 2–6, 2009.

[206] C.-H. Chen, D. S. Naidu, and M. P. Schoen. "An adaptive control strategy for a five-fingered prosthetic hand," in *The 14th World Scientific and Engineering Academy and Society (WSEAS) International Conference on Systems, Latest Trends on Systems (Volume II)*, pp. 405–410, Corfu Island, Greece, July 22–24, 2010.

[207] C.-H. Chen, M. P. Schoen, and K. W. Bosworth. "A condensed hybrid optimization algorithm using enhanced continuous tabu search and particle swarm optimization," in *Proceedings of the ASME 2009 Dynamic Systems and Control Conference (DSCC)*, Hollywood, California, USA, October 12–14, 2009 (No. DSCC2009–2526).

[208] F. Liu and H. Chen. "Motion control of intelligent underwater robot based on CMAC-PID," in *Proceedings of the 2008 IEEE International Conference on Information and Automation*, pp. 1308–1311, Zhangjiajie, P. R. China, June 20–23, 2008.

[209] L. A. Zadeh. Soft computing and fuzzy logic. *IEEE Software*, 11(6):48–56, 1994.

[210] L. A. Zadeh. Fuzzy sets. *Information and Control*, 8:338–353, 1965.

[211] L. A. Zadeh. Outline of a new approach to the analysis of complex systems and decision processes. *IEEE Transactions on Systems, Man, and Cybernetics*, 3(1):28–44, 1973.

[212] L. A. Zadeh. *Possibility Theory and Soft Data Analysis, book chapter 3, to appear in a book titled "Mathematical Frontiers of the Social and Policy Sciences"*. Westview Press, Boulder, Colorado, USA, 1981.

[213] L. Magdalena. What is soft computing? revisiting possible answers. *International Journal of Computational Intelligence Systems*, 3(2):148–159, June 2010.

[214] J.-S. R. Jang, C.-T. Sun, and E. Mizutani. *Neuro-Fuzzy and Soft Computing: A Computational Approach to Learning and Machine Intelligence*. Prentice Hall PTR, Upper Saddle River, New Jersey, USA, 1997.

[215] A. Tettamanzi and M. Tomassini. *Soft Computing: Integrating Evolutionary, Neural, and Fuzzy Systems*. Springer-Verlag, Berlin, Germany, 2001.

[216] S. J. Ovaska, H. F. VanLandingham, and A. Kamiya. Fusion of soft computing and hard computing in industrial applications: An overview. *IEEE Transactions on Systems, Man, and Cybernetics, Part C: Applications and Reviews*, 32(2):72–79, May 2002.

[217] F. O. Karray and C. De Silva. *Soft Computing and Intelligent Systems Design: Theory, Tools and Applications*. Pearson Educational Limited, Harlow, England, UK, 2004.

[218] A. Konar. *Computational Intelligence: Principles, Techniques and Applications*. Springer-Verlag, Berlin, Germany, 2005.

[219] C.-H. Chen, K. W. Bosworth, and M. P. Schoen. "Investigation of particle swarm optimization dynamics," in *Proceedings of International Mechanical Engineering Congress and Exposition (IMECE) 2007*, Seattle, Washington, USA, November 11–15, 2007 (No. IMECE2007–41343).

[220] C.-H. Chen, K. W. Bosworth, and M. P. Schoen. "An adaptive particle swarm method to multiple dimensional problems," in *Proceedings of the International Association of Science and Technology for Development (IASTED) International Symposium on Computational Biology and Bioinformatics (CBB 2008)*, pp. 260–265, Orlando, Florida, USA, November 16–18, 2008.

[221] C. I. Christodooulu and C. S. Pattichis. Unsupervised pattern recognition for the classification of EMG signals. *IEEE Transactions on Biomedical Engineering*, 46:169–178, 1999.

[222] F. H. Y. Chan, Y.-S. Yang, F. K. Lam, Y.-T. Zhang, and P. A. Parker. Fuzzy EMG classification for prosthesis control. *IEEE Transactions on Rehabilitation Engineering*, 8(3):305–311, 2000.

[223] J. J. Fernandez, K. A. Farry, and J. B. Cheatham. "Waveform recognition using genetic programming: The myoelectric signal recognition problem," in *Proceedings of the First Annual Conference of Genetic Programming*, pp. 1754–1759, 2000.

[224] K. Kim and J. E. Colgate. Haptic feedback enhances grip force control of sEMG-controlled prosthetic hands in targeted reinnervation amputees. *IEEE Transactions on Neural Systems and Rehabilitation Engineering*, 20(6):798–805, 2012.

[225] E. N. Kamavuakoa, J. C. Rosenvanga, M. F. Bga, A. Smidstrupa, E. Erkocevica, M. J. Niemeiera, W. Jensena, and D. Farinaa. Influence of the feature space on the estimation of hand grasping force from intramuscular EMG. *Biomedical Signal Processing and Control*, 8:1–5, 2013.

[226] E. D. Engeberg. A physiological basis for control of a prosthetic hand. *Biomedical Signal Processing and Control*, 8:6–15, 2013.

[227] D. S. Naidu. Intelligent Control Systems. Graduate Course Class Notes, 2007.

[228] J. T. Bingham and M. P. Schoen. "Characterization of myoelectric signals using system identification techniques," in *Proceedings of the 2004 ASME International Mechanical Engineering Congress and Exposition (IMECE)*, pp. 123–128, Anaheim, California, USA, November 13–19, 2004.

[229] K. Duraisamy, O. Isebor, A. Perez, M. P. Schoen, and D. S. Naidu. "Kinematic synthesis for smart hand prosthesis," in *Proceedings of the First IEEE/RAS-EMBS 2006 International Conference on Biomedical Robotics and Biomechatronics*, pp. 1135–1140, Pisa, Italy, February 20–22, 2006.

[230] D. S. Naidu and V. K. Nandikolla. "Fusion of hard and soft control strategies for left ventricular ejection dynamics arising in biomedicine," in *Proceedings of the Automatic Control Conference (ACC)*, pp. 1575–1580, Portland, Oregon, USA, June 8–10, 2005.

[231] V. K. Nandikolla and D. S. Naidu. "Blood glucose regulation for diabetic mellitus using a hybrid intelligent technique," in *Proceedings of the 2005 ASME International Mechanical Engineering Congress and Exposition (IMECE)*, pp. 1–6, Orlando, Florida, USA, November 5–11, 2005.

[232] S. J. Ovaska and H. F. VanLandingham. Guest editorial special issue on fusion of soft computing and hard computing in industrial applications. *IEEE Transaction on Systems, Man, and Cybernetics, Part C: Applications and Reviews*, 32(2):69–71, May 2002.

[233] B. Sick and S. J. Ovaska. "Fusion of soft and hard computing techniques: A multi-dimensional categorization scheme," in *2005 IEEE Mid-Summer Workshop on Soft Computing in Industrial Applications*, pp. 57–62, Espoo, Finland, June 28–30, 2005.

[234] S. J. Ovaska, A. Kamiya, and Y. Chen. Fusion of soft computing and hard computing: computational structures and characteristic features. *IEEE Transaction on Systems, Man, and Cybernetics, Part C: Applications and Reviews*, 36(3):439–448, May 2006.

[235] J. C. K. Lai, M. P. Schoen, A. Perez-Gracia, D. S. Naidu, and S. W. Leung. Prosthetic devices: Challenges and implications of robotic implants and biological interfaces. *Proceedings of the Institute of Mechanical Engineers (IMechE), Part H: Journal of Engineering in Medicine*, 221(2):173–183, January 2007 (Special Issue on Micro and Nano Technologies in Medicine: This article listed as 1 of 20 in *Top 20 Articles, in the Domain of Article 17385571, Since its Publication (2007)* according to *BioMedLib: "Who is Publishing in My Domain?"* as on March 17, 2015).

[236] C.-H. Chen, D. S. Naidu, and M. P. Schoen. Adaptive control for a five-fingered prosthetic hand with unknown mass and inertia. *World Scientific and Engineering Academy and Society (WSEAS) Journal on Systems*, 10(5):148–161, May 2011.

[237] C.-H. Chen and D. S. Naidu. "Fusion of fuzzy logic and PD control for a five-fingered smart prosthetic hand," in *Proceedings of the 2011 IEEE International Conference on Fuzzy Systems (FUZZ–IEEE 2011)*, pp. 2108–2115, Taipei, Taiwan, June 27–30, 2011.

[238] C.-H. Chen and D. S. Naidu. Hybrid control strategies for a five-finger robotic hand. *Biomedical Signal Processing and Control*, 8(4):382–390, July 2013.

[239] C.-H. Chen and D. S. Naidu. A modified optimal control strategy for a five-finger robotic hand. *International Journal of Robotics and Automation Technology*, 1(1):3–10, November 2014.

[240] C.-H. Chen and D. S. Naidu. "Hybrid genetic algorithm PID control for a five-fingered smart prosthetic hand," in *Proceedings of the Sixth International Conference on Circuits, Systems and Signals (CSS'11)*, pp. 57–63, Vouliagmeni Beach, Athens, Greece, March 7–9, 2012.

[241] K. W. Horch and G. S. Dhillon. *Neuroprosthesis: Theory and Practice*. World Scientific, River Edge, New Jersey, USA, 2004.

[242] X. Navarro, T. B. Krueger, N. Lago, S. Micera, T. Stieglitz, and P. Dario. A critical review of interfaces with the peripheral nervous system for the control of neuroprostheses and hybrid bionic systems. *Journal of the Peripheral Nervous System*, 10:229–258, 2005 (Review article with over 300 references).

[243] K. W. Horch and G. S. Dhillon. "Towards a neuroprosthetic arm," in *Proceedings of the First IEEE/RAS-EMBS 2006 International Conference on Biomedical Robotics and Biomechanics*, pp. 1–4, Pisa, Italy, February 2006.

[244] K. Shenoy. Toward high-performance neural control of prosthetic devices. Technical report, Stanford University, Stanford, California, USA, May 2007.

[245] R. L. Hart, K. L. Kilgore, and P. H. Peckham. A comparison between control methods for implanted FES hand-grasp systems. *IEEE Transactions on Rehabilitation Engineering*, 6(2):208–218, June 1998.

[246] K. Oppenheim. Jess Sullivan powers robotic arms with his mind. CNN, March 23, 2006.

[247] CNN-News: Bionic arm provides hope for amputees, September 14, 2006.

[248] P. Guinnessy. DARPA joins industry, academia to build better prosthetic arms. *Physics Today*, 59(9):24–25, September 2006.

[249] CNN-News: Brain chip heralds neurotech dawn, July 17, 2006.

[250] L. R. Hochberg, M. D. Serruya, G. M. Friehs, A. Mukand, M. Saleh, A. H. Caplan, A. Branner, D. Chen, R. D. Penn, and J. P. Donoghue. Neuronal ensemble control of prosthetic devices by a human with tetraplegia. *Nature*, 442(13):164–171, July 2006.

[251] N. Thakor. Catching brain waves in a net. *Spectrum, IEEE*, 51(9):40–45, September 2014.

CHAPTER 2

KINEMATICS AND TRAJECTORY PLANNING

This chapter addresses human hand anatomy, the problems of forward kinematics, inverse kinematics, differential kinematics, and trajectory planning. A serial manipulator is defined as a series of connected joints. One end of the chain is fixed in the base frame while the other end is free and for our purposes, the end-effector is called the fingertip of fingers. The resulting motion of the serial links is obtained by composition of the elementary motions of each link with respect to the previous link. Hence, in order to manipulate an object in space, the fingertip (end-effector) position of each finger must be described. Section 2.2 shows the derivation the fingertip positions by forward kinematics. Using inverse kinematics in Section 2.3, the joint angles of each finger (*joint space*) need to be obtained from the known fingertip positions (*Cartesian space*). In real life, the joint angle positions of each finger are constrained in the angular range. In Section 2.3, we generate the workspace of the fingertip. The linear and angular velocities and accelerations of fingertips are obtained by differential kinematics in Section 2.4. Then the joint angular velocities and joint angular accelerations of each finger are derived from the linear and angular velocities and accelerations of fingertips by the *geometric Jacobian*. Finally, before the robotic hand is controlled to execute a specific hand motion task, the desired paths are designed by polynomial and Bézier curve functions in Section 2.5.

Fusion of Hard and Soft Control Strategies for the Robotic Hand, By C.-H. Chen and D. S. Naidu   **47**
© 2017 by the Institute of Electrical and Electronic Engineers, Inc. Published 2017 by John Wiley & Sons, Inc.

2.1   Human Hand Anatomy

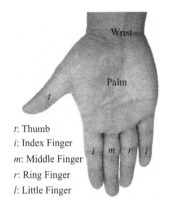

t: Thumb
i: Index Finger
m: Middle Finger
r: Ring Finger
l: Little Finger

(a) Physical Appearance of Right Hand
(Anterior View)

Carpals (proximal row)	S: Scaphoid bone	T: Trapezium bone	Carpals (distal row)
	L: Lunate bone	T: Trapezoid bone	
	T: Triquetrum bone	C: Capitate bone	
	P: Pisiform bone	H: Hamate bone	

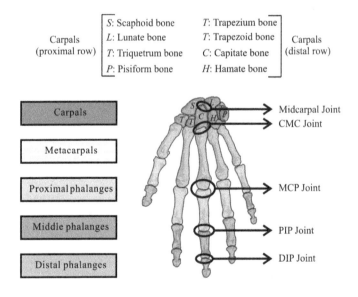

(b) Bones of Left Hand (Posterior View)

Figure 2.1   Human Wrist and Hand: (a) Physical Appearance of Right Hand (Anterior View): A Human Hand Has Thumb, Index, Middle, Ring, and Little Fingers, Palm, and Wrist. (b) Bones of Left Hand (Posterior View).

Figure 2.1(a) shows a normal human hand composed of thumb (t), index (i), middle (m), ring (r), little (l) fingers, and palm. The wrist is located between the forearm and the hand and consists of eight carpal bones organized in two rows of proximal (movable) and distal (immovable) carpal bones as shown in Figure 2.1(b) [1–3]. A human hand has 27 bones, including 5 distal phalanges, 4 middle phalanges, 5 proximal phalanges, 5 metacarpals, and 8 carpals. The proximal row (top) of carpal bones from lateral to medial is the scaphoid, lunate, triquetrum and pisiform; the distal row (bottom) of carpal bones from medial to lateral has the hamate, capitate, trapezoid and trapezium. The hand is composed of five metacarpals and five digits. The metacarpals produce a curve, so the palm is concave in the resting position. The five digits contain one thumb (t) and four fingers, for example, index (i), middle (m), ring (r), and little (l) fingers, respectively. The thumb has two bones, proximal phalanx and distal phalanx. Each finger consists of three bones, proximal phalanx, middle phalanx, and distal phalanx. In this work, we assumed that the palm is fixed, the thumb has two links (proximal phalanx and distal phalanx), and each finger has three links (proximal phalanx, middle phalanx, and distal phalanx).

Synovial joints are formed at the surface of relative motion between two bones. The joints of thumb and four fingers contain two saddle-shaped articulating surfaces between two bones and is4 classified as saddle joints. Index, middle, ring, and little fingers include three revolute joints in order to do the angular movements (Figure 2.1 (b)). Metacarpal-phalangeal (MCP) joint is located between metacarpal and proximal phalange bones; proximal and distal interphalangeal (PIP and DIP) joints separate the phalangeal bones. Thumb contains MCP and interphalangeal (IP) joints [1]. For a human hand, each finger has four DOFs (two at MCP joint, one at PIP joint, and one at DIP joint), thumb has three DOFs (two at MCP joint and one at IP joint), wrist has two DOFs and carpometacarpal (CMC) joint has two DOFs. In this work, we model 14-DOF, five-fingered robotic hand with two-link thumb and remaining three-link fingers. q_1^j, q_2^j, and q_3^j ($j = i$, m, r, and l) represent the angular positions (or joint angles) of the first joint MCP^j, the second joint PIP^j, and the third joint DIP^j of index, middle, ring, and little fingers, respectively; q_1^t and q_2^t are the angular positions of the first joint MCP^t and the second joint IP^t of thumb (t), respectively.

2.2 Forward Kinematics

Kinematics is the study of geometry in motion. It is restricted to a natural geometrical description of motion, including positions, orientations, and their derivatives (velocities and accelerations). In other words, forward and inverse kinematics of articulated systems study the analytical relationship between the angular positions of joints and the positions and orientations of the end-effector (fingertip). Differential kinematics then expresses the analytical relationship between the angular velocities and angular accelerations of joints and the linear and angular velocities and accelerations of end-effector (fingertip) by the geometric Jacobian of the manipulator. For robotic hand, the kinematic descriptions of manipulators are used to derive the fundamental equations for dynamics and control purposes. The coming three Sections 2.2–2.4 will introduce forward kinematics, inverse kinematics, and differential kinematics by means of a serial n-link manipulator, two-link thumb, three-link fingers, and three-dimensional five-fingered robotic hand, respectively.

2.2.1 Homogeneous Transformations

Before the forward kinematics is derived, it is necessary to study the rotation matrices, translation vectors, and homogeneous transformations [4]. Figure 2.2 shows

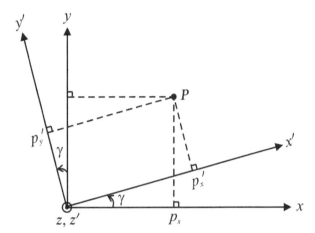

Figure 2.2   Representation of a Point P in a Rotation of Frames by an Angle γ about z Axis

two frames x-y-z and x'-y'-z' with the same origin mutually rotated by an angle γ about the z axis. Let $\mathbf{p} = [p_x \ p_y \ p_z]'$ and $\mathbf{p}' = [p'_x \ p'_y \ p'_z]'$ be the vectors of the coordinates of a point P in the two frames x-y-z and x'-y'-z', respectively. According to the geometry, the relationship between the two vectors of the coordinates of the point P in the two frames is expressed as

$$
\begin{aligned}
p_x &= p_x'\cos\gamma - p_y'\sin\gamma, \\
p_y &= p_x'\sin\gamma + p_y'\cos\gamma, \\
p_z &= p_z'.
\end{aligned}
\tag{2.2.1}
$$

(2.2.1) is expressed in the matrix form as follows

$$
\begin{bmatrix} p_x \\ p_y \\ p_z \end{bmatrix}
=
\begin{bmatrix} \cos\gamma & -\sin\gamma & 0 \\ \sin\gamma & \cos\gamma & 0 \\ 0 & 0 & 1 \end{bmatrix}
\begin{bmatrix} p_x' \\ p_y' \\ p_z' \end{bmatrix},
$$

$$
\mathbf{p} = \mathbf{R}_z(\gamma)\,\mathbf{p}'.
\tag{2.2.2}
$$

Here, the rotation matrix $\mathbf{R}_z(\gamma)$ represents the rotation of the frame x'-y'-z' with respect to the frame x-y-z by an angle γ about the z axis and is written as

$$
\mathbf{R}_z(\gamma) =
\begin{bmatrix} \cos\gamma & -\sin\gamma & 0 \\ \sin\gamma & \cos\gamma & 0 \\ 0 & 0 & 1 \end{bmatrix}.
\tag{2.2.3}
$$

In a similar method, it is demonstrated that the rotation matrices $\mathbf{R}_x(\alpha)$ and $\mathbf{R}_y(\beta)$ are, respectively, the rotation of the frame $x'\text{-}y'\text{-}z'$ with respect to the frame $x\text{-}y\text{-}z$ by angles α and β about x and y axes and are derived from

$$\mathbf{R}_x(\alpha) = \begin{bmatrix} 1 & 0 & 0 \\ 0 & \cos\alpha & -\sin\alpha \\ 0 & \sin\alpha & \cos\alpha \end{bmatrix}, \qquad (2.2.4)$$

$$\mathbf{R}_y(\beta) = \begin{bmatrix} \cos\beta & 0 & \sin\beta \\ 0 & 1 & 0 \\ -\sin\beta & 0 & \cos\beta \end{bmatrix}. \qquad (2.2.5)$$

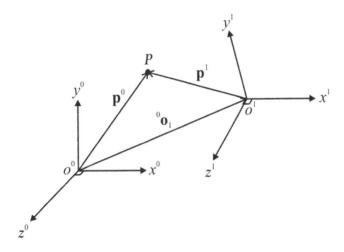

Figure 2.3 Representation of a Point P in Two Different Coordinates

Figure 2.3 shows an arbitrary point P in space. The position vectors of this point P in two different coordinate frames $o^0\text{-}x^0y^0z^0$ and $o^1\text{-}x^1y^1z^1$ are described as $\mathbf{p}^0$ and $\mathbf{p}^1$, respectively. Let $^0\mathbf{o}_1$ and $^0\mathbf{R}_1$ be the *translation* vector of the origin of the frame and *rotation* matrix of coordinate frame 1 with respect to coordinate frame 0, respectively. Hence, the position vector $\mathbf{p}^0$ of the point P with respect to coordinate frame $o^0\text{-}x^0y^0z^0$ is written as

$$\mathbf{p}^0 = {}^0\mathbf{o}_1 + {}^0\mathbf{R}_1\,\mathbf{p}^1. \qquad (2.2.6)$$

(2.2.6) is the coordinate transformation of translation vector $^0\mathbf{o}_1$ and rotation matrix $^0\mathbf{R}_1$ in two frames. The homogeneous transformation matrix $^0\mathbf{T}_1$ of coordinate frame 1 with respect to coordinate frame 0 is expressed in terms of the 4×4 matrix as

$$
{}^0\mathbf{T}_1 = \left[\begin{array}{c|c}
{}^0\mathbf{R}_1(3 \times 3) & {}^0\mathbf{o}_1(3 \times 1) \\
\hline
\mathbf{0}\ (1 \times 3) & 1\ (1 \times 1)
\end{array}\right], \tag{2.2.7}
$$

$$
= \left[\begin{array}{c|c}
\textbf{Rotation} & \textbf{Translation} \\
\hline
\textbf{Perspective} & \textbf{Scale Factor}
\end{array}\right].
$$

Therefore, the homogeneous position vectors of the point P with respect to two different frames 0 and 1 are written in terms of the homogeneous transformation matrix as

$$
\widetilde{\mathbf{p}}^0 = {}^0\mathbf{T}_1\ \widetilde{\mathbf{p}}^1. \tag{2.2.8}
$$

Here, $\widetilde{\mathbf{p}}^0 = [\mathbf{p}^0\ 1]'$ and $\widetilde{\mathbf{p}}^1 = [\mathbf{p}^1\ 1]'$. Prime $(')$ means the transpose of vectors or matrices.

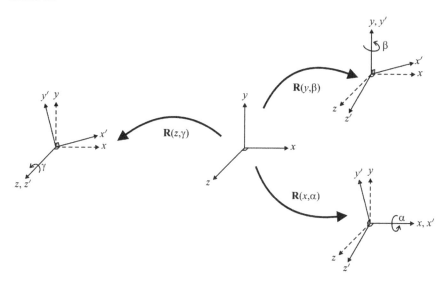

Figure 2.4   Homogeneous Transformations $\mathbf{R}(x, \alpha)$, $\mathbf{R}(y, \beta)$, and $\mathbf{R}(z, \gamma)$ of Rotation Matrices with Rotating Angles α, β, and γ about x, y, and z Axes

Figure 2.4 shows homogeneous transformations $\mathbf{R}(x, \alpha), \mathbf{R}(y, \beta)$, and $\mathbf{R}(z, \gamma)$ of rotation matrices with rotating angles α, β, and γ about x, y, and z axes, respectively. $\mathbf{R}(x, \alpha), \mathbf{R}(y, \beta)$, and $\mathbf{R}(z, \gamma)$ are written as

$$
\mathbf{R}(x, \alpha) = \begin{bmatrix}
1 & 0 & 0 & 0 \\
0 & \cos\alpha & -\sin\alpha & 0 \\
0 & \sin\alpha & \cos\alpha & 0 \\
0 & 0 & 0 & 1
\end{bmatrix},
$$

$$\mathbf{R}(y,\beta) = \begin{bmatrix} \cos\beta & 0 & \sin\beta & 0 \\ 0 & 1 & 0 & 0 \\ -\sin\beta & 0 & \cos\beta & 0 \\ 0 & 0 & 0 & 1 \end{bmatrix},$$

$$\mathbf{R}(z,\gamma) = \begin{bmatrix} \cos\gamma & -\sin\gamma & 0 & 0 \\ \sin\gamma & \cos\gamma & 0 & 0 \\ 0 & 0 & 1 & 0 \\ 0 & 0 & 0 & 1 \end{bmatrix}. \tag{2.2.9}$$

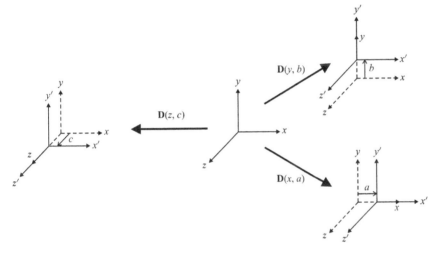

Figure 2.5 Homogeneous Transformations $\mathbf{D}(x,a)$, $\mathbf{D}(y,b)$, and $\mathbf{D}(z,c)$ of Translation Vectors with Displacements a, b, and c about x, y, and z Axes

Similarly, Figure 2.5 shows homogeneous transformations $\mathbf{D}(x,a)$, $\mathbf{D}(y,b)$, and $\mathbf{D}(z,c)$ of translation vectors with displacements a, b, and c about x, y, and z axes, respectively. $\mathbf{D}(x,a)$, $\mathbf{D}(y,b)$, and $\mathbf{D}(z,c)$ are written as

$$\mathbf{D}(x,a) = \begin{bmatrix} 1 & 0 & 0 & a \\ 0 & 1 & 0 & 0 \\ 0 & 0 & 1 & 0 \\ 0 & 0 & 0 & 1 \end{bmatrix},$$

$$\mathbf{D}(y,b) = \begin{bmatrix} 1 & 0 & 0 & 0 \\ 0 & 1 & 0 & b \\ 0 & 0 & 1 & 0 \\ 0 & 0 & 0 & 1 \end{bmatrix},$$

$$\mathbf{D}(z, c) = \begin{bmatrix} 1 & 0 & 0 & 0 \\ 0 & 1 & 0 & 0 \\ 0 & 0 & 1 & c \\ 0 & 0 & 0 & 1 \end{bmatrix}. \qquad (2.2.10)$$

2.2.2 Serial n Link Revolute Joint Planar Manipulator

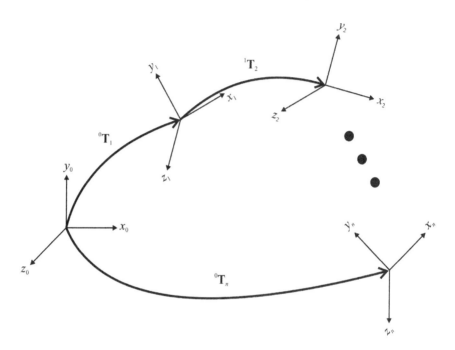

Figure 2.6 Homogeneous Transformation Matrix $^0\mathbf{T}_n$ of Coordinate Frame n with Respect to Coordinate Frame 0

As shown in Figure 2.6, homogeneous transformation matrix $^0\mathbf{T}_n$ of coordinate frame n with respect to coordinate frame 0 is considered as the composition of transformations along the serial n frames and $^0\mathbf{T}_n$ is derived from

$$^0\mathbf{T}_n = {}^0\mathbf{T}_1 {}^1\mathbf{T}_2 \cdots {}^{n-1}\mathbf{T}_n. \qquad (2.2.11)$$

Based on the recursive expressions in (2.2.11), a general systematic method needs to be derived to define the relative position and orientation of two consecutive links in order to calculate the forward kinematic equations for a serial n-link revolute-joint manipulator [5]. Accordingly, we need to compute the coordinate transformation matrix $^{i-1}\mathbf{T}_i$ between the frame i attached to the $(i-1)$th and ith links.

Figure 2.7 shows the illustration of a serial n-link revolute-joint planar manipulator. L_i is the length of the link i and q_i is the angle of the joint i ($i = 1, 2, \cdots, n$). The *Denavit–Hartenberg* (DH) method [6–10] is used to define the position

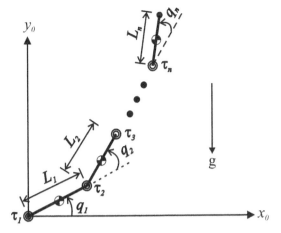

Figure 2.7   Illustration of a Serial n-Link Revolute-Joint Planar Manipulator

of frame i with respect to the previous $i - 1$ frame. The DH coordinate frame is identified by four parameters: a_i, α_i, d_i, and θ_i. a_i is the kinematic length of the link i; α_i is the twist angle of the link i; d_i means the link offset (or called joint distance), which is the distance between two joint axes; θ_i represents the joint angle. A convenient zero configuration is to consider all links extended along the x axis. Consequently, according to the DH convention, the coordinate transformation matrix $^{i-1}\mathbf{T}_i$ to transform coordinate frame i with respect to frame $i - 1$ is expressed as a product of four basic homogeneous transformations $\mathbf{D}(z_{i-1}, d_i)$, $\mathbf{R}(z_{i-1}, \theta_i)$, $\mathbf{D}(x_{i-1}, a_i)$, and $\mathbf{R}(x_{i-1}, \alpha_i)$.

$$
\begin{aligned}
^{i-1}\mathbf{T}_i &= \mathbf{D}(z_{i-1}, d_i)\,\mathbf{R}(z_{i-1}, \theta_i)\,\mathbf{D}(x_{i-1}, a_i)\,\mathbf{R}(x_{i-1}, \alpha_i), \\
&= \begin{bmatrix}
\cos\theta_i & -\sin\theta_i\cos\alpha_i & \sin\theta_i\sin\alpha_i & a_i\cos\theta_i \\
\sin\theta_i & \cos\theta_i\cos\alpha_i & -\cos\theta_i\sin\alpha_i & a_i\sin\theta_i \\
0 & \sin\alpha_i & \cos\alpha_i & d_i \\
0 & 0 & 0 & 1
\end{bmatrix}, \quad (2.2.12)
\end{aligned}
$$

where

$$
\mathbf{D}(z_{i-1}, d_i) = \begin{bmatrix}
1 & 0 & 0 & 0 \\
0 & 1 & 0 & 0 \\
0 & 0 & 1 & d_i \\
0 & 0 & 0 & 1
\end{bmatrix}, \quad
\mathbf{R}(z_{i-1}, \theta_i) = \begin{bmatrix}
\cos\theta_i & -\sin\theta_i & 0 & 0 \\
\sin\theta_i & \cos\theta_i & 0 & 0 \\
0 & 0 & 1 & 0 \\
0 & 0 & 0 & 1
\end{bmatrix},
$$

$$
\mathbf{D}(x_{i-1}, a_i) \;=\; \begin{bmatrix} 1 & 0 & 0 & a_i \\ 0 & 1 & 0 & 0 \\ 0 & 0 & 1 & 0 \\ 0 & 0 & 0 & 1 \end{bmatrix}, \quad \mathbf{R}(x_{i-1}, \alpha_i) = \begin{bmatrix} 1 & 0 & 0 & 0 \\ 0 & \cos\alpha_i & -\sin\alpha_i & 0 \\ 0 & \sin\alpha_i & \cos\alpha_i & 0 \\ 0 & 0 & 0 & 1 \end{bmatrix}.
$$

$$(2.2.13)$$

Similar to (2.2.8), the transformation equation from coordinate frame $i = [x_i \; y_i \; z_i]'$ to its previous coordinate frame $i - 1 = [x_{i-1} \; y_{i-1} \; z_{i-1}]'$ is

$$
\begin{bmatrix} x_{i-1} \\ y_{i-1} \\ z_{i-1} \\ 1 \end{bmatrix} = {}^{i-1}\mathbf{T}_i \begin{bmatrix} x_i \\ y_i \\ z_i \\ 1 \end{bmatrix}.
$$

$$(2.2.14)$$

The transformation matrix ${}^{i}\mathbf{T}_{i-1}$ from previous coordinate frame $i - 1 = [x_{i-1} \; y_{i-1} \; z_{i-1}]'$ to coordinate frame $i = [x_i \; y_i \; z_i]'$ is obtained from the inverse of ${}^{i-1}\mathbf{T}_i$.

$$
\begin{bmatrix} x_i \\ y_i \\ z_i \\ 1 \end{bmatrix} = {}^{i}\mathbf{T}_{i-1} \begin{bmatrix} x_{i-1} \\ y_{i-1} \\ z_{i-1} \\ 1 \end{bmatrix},
$$

$$(2.2.15)$$

where

$$
\begin{aligned}
{}^{i}\mathbf{T}_{i-1} &= {}^{i-1}\mathbf{T}_i^{-1}, \\
&= \begin{bmatrix} \cos\theta_i & \sin\theta_i & 0 & -a_i \\ -\sin\theta_i\cos\alpha_i & \cos\theta_i\cos\alpha_i & \sin\alpha_i & -d_i\sin\alpha_i \\ \sin\theta_i\sin\alpha_i & -\cos\theta_i\sin\alpha_i & \cos\alpha_i & -d_i\cos\alpha_i \\ 0 & 0 & 0 & 1 \end{bmatrix}
\end{aligned}
$$

$$(2.2.16)$$

Table 2.1 is a DH parameter table for an n-link revolute-joint planar manipulator shown in Figure 2.7. Therefore, using these parameters in (2.2.12), the coordinate transformation matrix ${}^{i-1}\mathbf{T}_i$ to transform coordinate frame i to frame $i - 1$ is computed by

$$
{}^{i-1}\mathbf{T}_i \;=\; \begin{bmatrix} \cos q_i & -\sin q_i & 0 & L_i \cos q_i \\ \sin q_i & \cos q_i & 0 & L_i \sin q_i \\ 0 & 0 & 1 & 0 \\ 0 & 0 & 0 & 1 \end{bmatrix}.
$$

$$(2.2.17)$$

Table 2.1   DH Parameter Table for an n-Link Planar Manipulator Shown in Figure 2.7

Link No.	a_i	α_i	d_i	θ_i
0	0	0	0	0
1	L_1	0	0	q_1
2	L_2	0	0	q_2
$\vdots$	$\vdots$	$\vdots$	$\vdots$	$\vdots$
i	L_i	0	0	q_i
$\vdots$	$\vdots$	$\vdots$	$\vdots$	$\vdots$
n	L_n	0	0	q_n

Consequently, taking (2.2.17) into (2.2.11), (2.2.12) is rewritten as

$$
\begin{aligned}
{}^{0}\mathbf{T}_n &= {}^{0}\mathbf{T}_1{}^{1}\mathbf{T}_2\cdots{}^{i-1}\mathbf{T}_i\cdots{}^{n-1}\mathbf{T}_n, \\[6pt]
&= \begin{bmatrix}
C_{12\ldots n} & -S_{12\ldots n} & 0 & \sum_{i=1}^{n} L_i C_{12\ldots i} \\
S_{12\ldots n} & C_{12\ldots n} & 0 & \sum_{i=1}^{n} L_i S_{12\ldots i} \\
0 & 0 & 1 & 0 \\
0 & 0 & 0 & 1
\end{bmatrix},
\end{aligned} \tag{2.2.18}
$$

where

$$
{}^{0}\mathbf{T}_1 = \begin{bmatrix}
\cos q_1 & -\sin q_1 & 0 & L_1 \cos q_1 \\
\sin q_1 & \cos q_1 & 0 & L_1 \sin q_1 \\
0 & 0 & 1 & 0 \\
0 & 0 & 0 & 1
\end{bmatrix},
$$

$$
{}^{1}\mathbf{T}_2 = \begin{bmatrix}
\cos q_2 & -\sin q_2 & 0 & L_2 \cos q_2 \\
\sin q_2 & \cos q_2 & 0 & L_2 \sin q_2 \\
0 & 0 & 1 & 0 \\
0 & 0 & 0 & 1
\end{bmatrix},
$$

$$
{}^{n-1}\mathbf{T}_n = \begin{bmatrix}
\cos q_n & -\sin q_n & 0 & L_n \cos q_n \\
\sin q_n & \cos q_n & 0 & L_n \sin q_n \\
0 & 0 & 1 & 0 \\
0 & 0 & 0 & 1
\end{bmatrix}. \tag{2.2.19}
$$

$$S_{i...j} = \sin(q_i + \cdots + q_j),$$
$$C_{i...j} = \cos(q_i + \cdots + q_j). \tag{2.2.20}$$

Thus, the position P_n (X_n, Y_n) of the end-effector and the orientation ϕ_n of the end-effector frame is obtained from

$$X_n = \sum_{i=1}^{n} L_i C_{12...i} = \sum_{j=1}^{n} L_j \cos\left(\sum_{i=1}^{j} q_i\right),$$

$$Y_n = \sum_{i=1}^{n} L_i S_{12...i} = \sum_{j=1}^{n} L_j \sin\left(\sum_{i=1}^{j} q_i\right),$$

$$\phi_n = \sum_{i=1}^{n} q_i. \tag{2.2.21}$$

2.2.3 Two Link Thumb

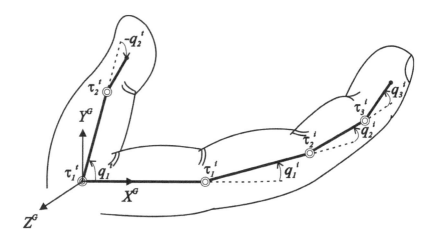

Figure 2.8   Schematic Diagram of Thumb and Index Finger

The links in kinematics are modeled as *rigid bodies*, so the properties of rigid body displacement take a central place in kinematics [9]. As shown in Figure 2.8, thumb (t) is assumed as two-link finger and the other four fingers, including index finger (i), middle finger (m), ring finger (r), and little finger (l), are considered as three-link fingers (Section 2.1).

Figure 2.9 shows the illustration of two-link thumb. L_1^t and L_2^t are the lengths of the links 1 and 2 of the thumb (t), respectively; q_1^t and q_2^t are the angles of joints 1 and 2 of the thumb [11]. Using DH method [6, 7, 9, 10], the fingertip (end-effector) coordinate $\mathbf{P}^t$ (X^t, Y^t) of the thumb is obtained by DH transformation matrices.

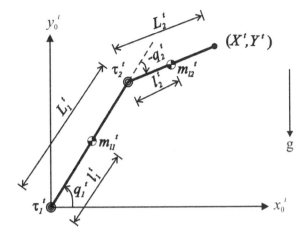

Figure 2.9 Two-Link Thumb Illustration

Table 2.2 DH Parameter Table for Two-Link Thumb Shown in Figure 2.9

Link No.	a_i	α_i	d_i	θ_i
0	0	0	0	0
1	L_1^t	0	0	q_1^t
2	L_2^t	0	0	q_2^t

Table 2.2 is the DH parameter table for two-link thumb shown in Figure 2.9. According to Table 2.2, the transformation matrices $^G\mathbf{T}_0^t$, $^0\mathbf{T}_1^t$ and $^1\mathbf{T}_2^t$ are found

$$
^G\mathbf{T}_0^t = \begin{bmatrix} 1 & 0 & 0 & 0 \\ 0 & 1 & 0 & 0 \\ 0 & 0 & 1 & 0 \\ 0 & 0 & 0 & 1 \end{bmatrix},
$$

$$
^0\mathbf{T}_1^t = \begin{bmatrix} \cos(q_1^t) & -\sin(q_1^t) & 0 & L_1^t\cos(q_1^t) \\ \sin(q_1^t) & \cos(q_1^t) & 0 & L_1^t\sin(q_1^t) \\ 0 & 0 & 1 & 0 \\ 0 & 0 & 0 & 1 \end{bmatrix},
$$

$$
^1\mathbf{T}_2^t = \begin{bmatrix} \cos(q_2^t) & -\sin(q_2^t) & 0 & L_2^t\cos(q_2^t) \\ \sin(q_2^t) & \cos(q_2^t) & 0 & L_2^t\sin(q_2^t) \\ 0 & 0 & 1 & 0 \\ 0 & 0 & 0 & 1 \end{bmatrix}. \tag{2.2.22}
$$

Here $^{G}\mathbf{T}_0^t$ is the transformation matrix from thumb local frame base (zero) to global (G) frame. $^{0}\mathbf{T}_1^t$ and $^{1}\mathbf{T}_2^t$ represent the transformation matrices from frame 1 to frame base (zero) and from frame 2 to frame 1, respectively. Consequently, the transformation matrix from thumb local frame 2 to global frame $^{G}\mathbf{T}_2^t$ is written as

$$
\begin{aligned}
^{G}\mathbf{T}_2^t &= \ ^{G}\mathbf{T}_0^t \ ^{0}\mathbf{T}_1^t \ ^{1}\mathbf{T}_2^t, \\
&= \begin{bmatrix} \cos(q_1^t + q_2^t) & -\sin(q_1^t + q_2^t) & 0 & L_1^t\cos(q_1^t) + L_2^t\cos(q_1^t + q_2^t) \\ \sin(q_1^t + q_2^t) & \cos(q_1^t + q_2^t) & 0 & L_1^t\sin(q_1^t) + L_2^t\sin(q_1^t + q_2^t) \\ 0 & 0 & 1 & 0 \\ 0 & 0 & 0 & 1 \end{bmatrix}.
\end{aligned}
$$
(2.2.23)

Thus, the fingertip coordinate $\mathbf{P}^t$ (X^t, Y^t) of the thumb (t) and the orientation ϕ^t of the fingertip frame are described as

$$
\begin{aligned}
X^t &= L_1^t\cos(q_1^t) + L_2^t\cos(q_1^t + q_2^t), \\
Y^t &= L_1^t\sin(q_1^t) + L_2^t\sin(q_1^t + q_2^t), \\
\phi^t &= q_1^t + q_2^t.
\end{aligned}
$$
(2.2.24)

2.2.4 Three Link Index Finger

Figure 2.10 shows the illustration of three-link index finger. d is the distance between

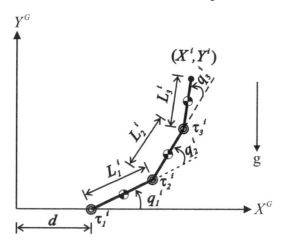

Figure 2.10 Three-Link Index Finger Illustration

global (G) frame and index finger (i) local frame base (zero); L_1^i, L_2^i, and L_3^i are the lengths of the links 1, 2, and 3 of the index finger (i), respectively; q_1^i, q_2^i, and q_3^i are the angles of the joints 1, 2, and 3 of the index finger [12]. Similarly, using DH method [6, 7, 9, 10], the fingertip (end-effector) coordinate $\mathbf{P}^i$ (X^i, Y^i) of the index finger is obtained by DH transformation matrices.

Table 2.3   DH Parameter Table for Three-Link Index Finger Shown in Figure 2.10

Link No.	a_i	α_i	d_i	θ_i
0	d	0	0	0
1	L_1^i	0	0	q_1^i
2	L_2^i	0	0	q_2^i
3	L_3^i	0	0	q_3^i

Table 2.3 is the DH parameter table for three-link index finger shown in Figure 2.10. Based on Table 2.3, the transformation matrices $^{G}\mathbf{T}_0^i$, $^{0}\mathbf{T}_1^i$, $^{1}\mathbf{T}_2^i$, and $^{2}\mathbf{T}_3^i$ are written as

$$
^{G}\mathbf{T}_0^i = \begin{bmatrix} 1 & 0 & 0 & d \\ 0 & 1 & 0 & 0 \\ 0 & 0 & 1 & 0 \\ 0 & 0 & 0 & 1 \end{bmatrix},
$$

$$
^{0}\mathbf{T}_1^i = \begin{bmatrix} \cos(q_1^i) & -\sin(q_1^i) & 0 & L_1^i \cos(q_1^i) \\ \sin(q_1^i) & \cos(q_1^i) & 0 & L_1^i \sin(q_1^i) \\ 0 & 0 & 1 & 0 \\ 0 & 0 & 0 & 1 \end{bmatrix},
$$

$$
^{1}\mathbf{T}_2^i = \begin{bmatrix} \cos(q_2^i) & -\sin(q_2^i) & 0 & L_2^i \cos(q_2^i) \\ \sin(q_2^i) & \cos(q_2^i) & 0 & L_2^i \sin(q_2^i) \\ 0 & 0 & 1 & 0 \\ 0 & 0 & 0 & 1 \end{bmatrix},
$$

$$
^{2}\mathbf{T}_3^i = \begin{bmatrix} \cos(q_3^i) & -\sin(q_3^i) & 0 & L_3^i \cos(q_3^i) \\ \sin(q_3^i) & \cos(q_3^i) & 0 & L_3^i \sin(q_3^i) \\ 0 & 0 & 1 & 0 \\ 0 & 0 & 0 & 1 \end{bmatrix}. \tag{2.2.25}
$$

Here, $^{G}\mathbf{T}_0^i$ is the transformation matrix from index finger (i) local frame base 0 to global (G) frame; $^{0}\mathbf{T}_1^i$, $^{1}\mathbf{T}_2^i$, and $^{2}\mathbf{T}_3^i$ are the transformation matrices from frame 1 to frame base, from frame 2 to frame 1, and from frame 3 to frame 2, respectively. As a result, the transformation matrix from index finger local frame 3 to global frame

$^{G}\mathbf{T}_3^i$ is written as

$$
^{G}\mathbf{T}_3^i = {}^{G}\mathbf{T}_0^i \, {}^{0}\mathbf{T}_1^i \, {}^{1}\mathbf{T}_2^i \, {}^{2}\mathbf{T}_3^i,
$$

$$
= \begin{bmatrix}
C_{123}^i & -S_{123}^i & 0 & d + L_1^i C_1^i + L_2^i C_{12}^i + L_3^i C_{123}^i \\
S_{123}^i & C_{123}^i & 0 & L_1^i S_1^i + L_2^i S_{12}^i + L_3^i S_{123}^i \\
0 & 0 & 1 & 0 \\
0 & 0 & 0 & 1
\end{bmatrix}, \quad (2.2.26)
$$

where, for simplicity, we have utilized the notations $C_1^i = \cos(q_1^i)$, $S_1^i = \sin(q_1^i)$, $C_{12}^i = \cos(q_1^i + q_2^i)$, $S_{12}^i = \sin(q_1^i + q_2^i)$, $C_{123}^i = \cos(q_1^i + q_2^i + q_3^i)$, and $S_{123}^i = \sin(q_1^i + q_2^i + q_3^i)$. Hence, the fingertip coordinate $\mathbf{P}^i$ (X^i, Y^i) and the orientation ϕ^i of the index finger (i) are written as

$$
\begin{aligned}
X^i &= d + L_1^i \cos(q_1^i) + L_2^i \cos(q_1^i + q_2^i) + L_3^i \cos(q_1^i + q_2^i + q_3^i), \\
Y^i &= L_1^i \sin(q_1^i) + L_2^i \sin(q_1^i + q_2^i) + L_3^i \sin(q_1^i + q_2^i + q_3^i), \\
\phi^i &= q_1^i + q_2^i + q_3^i. \quad (2.2.27)
\end{aligned}
$$

2.2.5   Three Dimensional Five Fingered Robotic Hand

As shown in Figure 2.11, index finger, middle finer, ring finer, and little finger include three revolute joints in order to do the angular movements. MCP joint is located between metacarpal and proximal phalange bone; PIP and DIP joints separate the phalangeal bones. Thumb contains MCP and IP joints (Section 2.1) [1]. In this book, q_1^j, q_2^j, and q_3^j represent the angular positions (or joint angles) of the first joint MCP^j, the second joint PIP^j, and the third joint DIP^j of index finger $(j = i)$, middle finger $(j = m)$, ring finger $(j = r)$, and little finger $(j = l)$, respectively; q_1^t and q_2^t are the angular positions of the first joint MCP^t and the second joint IP^t of thumb (t).

For a five-finger robotic hand shown in Figure 2.12, X^G, Y^G, and Z^G are the three axes of global coordinate. Local coordinate x^t-y^t-z^t of thumb is reached by rotating through angles α and β to X^G and Y^G of the global coordinate, subsequently. Local coordinate x^i-y^i-z^i of index finger is obtained by rotating through angle α to X^G and then translating a vector $\mathbf{d}^i$ of the global coordinate. Similarly, the local coordinate x^j-y^j-z^j of middle finger $(j = m)$, ring finger $(j = r)$, and little finger $(j = l)$ is obtained by rotating through angle α to X^G and then translating the vector $\mathbf{d}^j$ $(j = m, r, \text{and } l)$ of the global coordinate [3].

$$
\begin{aligned}
^{G}\mathbf{T}_t &= \mathbf{R}(X^G, \alpha) \, \mathbf{R}(Y^G, \beta), \\
&= \begin{bmatrix}
\cos \beta & 0 & \sin \beta & 0 \\
\sin \alpha \sin \beta & \cos \alpha & -\sin \alpha \cos \beta & 0 \\
-\cos \alpha \sin \beta & \sin \alpha & \cos \alpha \cos \beta & 0 \\
0 & 0 & 0 & 1
\end{bmatrix}, \quad (2.2.28)
\end{aligned}
$$

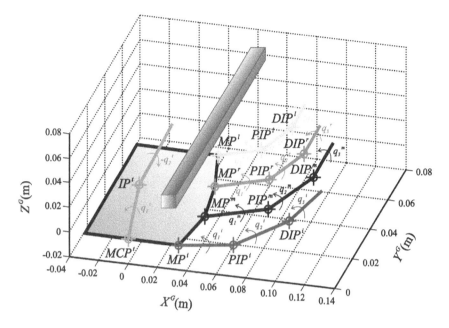

Figure 2.11   Joints of Five-Finger Robotic Hand Reaching a Rectangular Rod

where

$$
\mathbf{R}(X^G, \alpha) =
\begin{bmatrix}
1 & 0 & 0 & 0 \\
0 & \cos\alpha & -\sin\alpha & 0 \\
0 & \sin\alpha & \cos\alpha & 0 \\
0 & 0 & 0 & 1
\end{bmatrix}, \quad
\mathbf{R}(Y^G, \beta) =
\begin{bmatrix}
\cos\beta & 0 & \sin\beta & 0 \\
0 & 1 & 0 & 0 \\
-\sin\beta & 0 & \cos\beta & 0 \\
0 & 0 & 0 & 1
\end{bmatrix}.
$$

$$(2.2.29)$$

Let $\mathbf{p}^G = [p_X^G \ \ p_Y^G \ \ p_Z^G]'$ and $\mathbf{p}^t = [p_x^t \ \ p_y^t \ \ p_z^t]'$ be the position vectors of an arbitrary point P in the global coordinate $X^G\text{-}Y^G\text{-}Z^G$ and the thumb local frame base $x^t\text{-}y^t\text{-}z^t$, respectively. Therefore, $\tilde{\mathbf{p}}^G = [p_X^G \ \ p_Y^G \ \ p_Z^G \ \ 1]'$ is calculated by the

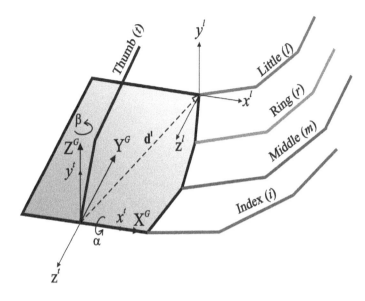

Figure 2.12   Relationship between Global Coordinate and Local Coordinates

product of $^G\mathbf{T}_t$ and $\tilde{\mathbf{p}}^t = [p_x^t \quad p_y^t \quad p_z^t \quad 1]'$.

$$
\begin{aligned}
\tilde{\mathbf{p}}^G &= {}^G\mathbf{T}_t\,\tilde{\mathbf{p}}^t, \\[4pt]
&= \begin{bmatrix}
\cos\beta & 0 & \sin\beta & 0 \\
\sin\alpha\,\sin\beta & \cos\alpha & -\sin\alpha\,\cos\beta & 0 \\
-\cos\alpha\,\sin\beta & \sin\alpha & \cos\alpha\,\cos\beta & 0 \\
0 & 0 & 0 & 1
\end{bmatrix}
\begin{bmatrix}
p_x^t \\ p_y^t \\ p_z^t \\ 1
\end{bmatrix}, \\[4pt]
&= \begin{bmatrix}
\cos\beta\,p_x^t + \sin\beta\,p_z^t \\
\sin\alpha\,\sin\beta\,p_x^t + \cos\alpha\,p_y^t - \sin\alpha\,\cos\beta\,p_z^t \\
-\cos\alpha\,\sin\beta\,p_x^t + \sin\alpha\,p_y^t + \cos\alpha\,\cos\beta\,p_z^t \\
1
\end{bmatrix}.
\end{aligned}
\tag{2.2.30}
$$

$\tilde{\mathbf{p}}^t$ is also computed by the product of $^G\mathbf{T}_t^{-1}$ and $\tilde{\mathbf{p}}^G$

$$
\tilde{\mathbf{p}}^t = {}^G\mathbf{T}_t^{-1}\,\tilde{\mathbf{p}}^G.
\tag{2.2.31}
$$

Figure 2.12 also shows that the local frame base x^i-y^i-z^i of index finger (i) is obtained by rotating through an angle α to X^G and then translating the vector $\mathbf{d}^j$ to the global frame X^G-Y^G-Z^G. Similarly, the local frame base x^j-y^j-z^j of middle finger $(j = m)$, ring finger $(j = r)$, and little finger $(j = l)$ is obtained by rotating through an angle α to X^G and then translating the vector $\mathbf{d}^j = [d_x^j \quad d_y^j \quad d_z^j]'$ $(j = m, r,$ and $l)$ with respect to the global frame. Therefore, the homogeneous transformation matrix $^G\mathbf{T}_j$ to transform the three-link finger local coordinate base frames x^j-y^j-z^j

$(j = i, m, r, \text{ and } l)$ to the global frame X^G-Y^G-Z^G is expressed as a product of four basic homogeneous transformations $\mathbf{R}(X^G, \alpha)$, $\mathbf{D}(X^G, d_x^j)$, $\mathbf{D}(Y^G, d_y^j)$, and $\mathbf{D}(Z^G, d_z^j)$.

$$
\begin{aligned}
{}^G\mathbf{T}_j &= \mathbf{R}(X^G, \alpha)\, \mathbf{D}(X^G, d_x^j)\, \mathbf{D}(Y^G, d_y^j)\, \mathbf{D}(Z^G, d_z^j), \\
&= \begin{bmatrix}
1 & 0 & 0 & d_x^j \\
0 & \cos\alpha & -\sin\alpha & d_y^j \\
0 & \sin\alpha & \cos\alpha & d_z^j \\
0 & 0 & 0 & 1
\end{bmatrix},
\end{aligned}
\tag{2.2.32}
$$

where

$$
\mathbf{D}(X^G, d_x^j) = \begin{bmatrix}
1 & 0 & 0 & d_x^j \\
0 & 1 & 0 & 0 \\
0 & 0 & 1 & 0 \\
0 & 0 & 0 & 1
\end{bmatrix}, \quad
\mathbf{D}(Y^G, d_y^j) = \begin{bmatrix}
1 & 0 & 0 & 0 \\
0 & 1 & 0 & d_y^j \\
0 & 0 & 1 & 0 \\
0 & 0 & 0 & 1
\end{bmatrix},
$$

$$
\mathbf{D}(Z^G, d_z^j) = \begin{bmatrix}
1 & 0 & 0 & 0 \\
0 & 1 & 0 & 0 \\
0 & 0 & 1 & d_z^j \\
0 & 0 & 0 & 1
\end{bmatrix}.
\tag{2.2.33}
$$

Let $\mathbf{p}^j = [p_x^j \ \ p_y^j \ \ p_z^j]'$ be the position vectors of an arbitrary point P in the three-link finger local frame base x_0^j-y_0^j-z_0^j ($j = i, m, r, \text{ and } l$). Therefore, $\mathbf{p}^G$ is computed by the product of ${}^G\mathbf{T}_j$ and $\tilde{\mathbf{p}}^j = [p_x^j \ \ p_y^j \ \ p_z^j \ \ 1]'$.

$$
\begin{aligned}
\tilde{\mathbf{p}}^G &= {}^G\mathbf{T}_j\, \tilde{\mathbf{p}}^j, \\
&= \begin{bmatrix}
1 & 0 & 0 & d_x^j \\
0 & \cos\alpha & -\sin\alpha & d_y^j \\
0 & \sin\alpha & \cos\alpha & d_z^j \\
0 & 0 & 0 & 1
\end{bmatrix}
\begin{bmatrix}
p_x^j \\
p_y^j \\
p_z^j \\
1
\end{bmatrix}, \\
&= \begin{bmatrix}
p_x^j + d_x^j \\
\cos\alpha\, p_y^j - \sin\alpha\, p_z^j + d_y^j \\
\sin\alpha\, p_y^j + \cos\alpha\, p_z^j + d_z^j \\
1
\end{bmatrix}.
\end{aligned}
\tag{2.2.34}
$$

$\tilde{\mathbf{p}}^j$ is also calculated by the product of ${}^G\mathbf{T}_j^{-1}$ and $\tilde{\mathbf{p}}^G$

$$
\tilde{\mathbf{p}}^j = {}^G\mathbf{T}_j^{-1}\, \tilde{\mathbf{p}}^G.
\tag{2.2.35}
$$

2.3 Inverse Kinematics

A desired trajectory (Section 2.5) is usually specified in *Cartesian space* and the trajectory controller is easily performed in the *joint space*. Hence, it is necessary to convert Cartesian trajectory planning to the joint space [6, 7, 9, 10]. Using inverse kinematics, the joint angular positions of each finger need to be obtained from the known fingertip positions (joint space). Then the angular velocities and angular accelerations of joints are obtained from the linear and angular velocities and accelerations of fingertips (end-effectors) by the *geometric Jacobian*.

2.3.1 Two Link Thumb

The joint angular positions of each finger is deduced as follows. According to forward kinematics [6, 7, 9, 10] (Section 2.2), the fingertip coordinate $\mathbf{P}^t$ (X^t, Y^t) of the thumb (t) (2.2.24) is described as

$$
\begin{aligned}
X^t &= L_1^t \cos(q_1^t) + L_2^t \cos(q_1^t + q_2^t), \\
Y^t &= L_1^t \sin(q_1^t) + L_2^t \sin(q_1^t + q_2^t).
\end{aligned}
\tag{2.3.1}
$$

Here L_1^t and L_2^t are the lengths of the links 1 and 2 of the thumb, respectively; q_1^t and q_2^t are the angular positions (or called angles) of joints 1 and 2 of the thumb. The sum of squared (2.3.1) is written as

$$
\begin{aligned}
X^{t2} + Y^{t2} &= L_1^{t\,2} \cos^2(q_1^t) + L_2^{t\,2} \cos^2(q_1^t + q_2^t) + 2L_1^t L_2^t \cos(q_1^t) \cos(q_1^t + q_2^t) \\
&\quad + L_1^{t\,2} \sin^2(q_1^t) + L_2^{t\,2} \sin^2(q_1^t + q_2^t) + 2L_1^t L_2^t \sin(q_1^t) \sin(q_1^t + q_2^t), \\
&= L_1^{t\,2} + L_2^{t\,2} + 2L_1^t L_2^t \cos(q_2^t).
\end{aligned}
\tag{2.3.2}
$$

Rearranging (2.3.2), we get the equation below.

$$
\cos(q_2^t) = \frac{X^{t2} + Y^{t2} - L_1^{t\,2} - L_2^{t\,2}}{2L_1^t L_2^t}.
\tag{2.3.3}
$$

Choosing the *elbow up* configuration, the angle q_2^t of the joint 2 is obtained from

$$
q_2^t = -\cos^{-1}\left(\frac{X^{t2} + Y^{t2} - L_1^{t\,2} - L_2^{t\,2}}{2L_1^t L_2^t} \right).
\tag{2.3.4}
$$

Notice that in this book, all positive angles are defined counterclockwise. When choosing the *elbow up* configuration, the angle q_2^t is clockwise, so the sign of q_2^t is negative.

Figure 2.13 is the geometric illustration of two-link thumb (elbow up) [2, 13]. Based on the geometry, we get two triangular relations below.

$$
\begin{aligned}
\tan\left(\alpha^t \right) &= \frac{Y^t}{X^t}, \\
\tan\left(\beta^t \right) &= -\frac{L_2^t \sin(q_2^t)}{L_1^t + L_2^t \cos\left(q_2^t\right)}.
\end{aligned}
\tag{2.3.5}
$$

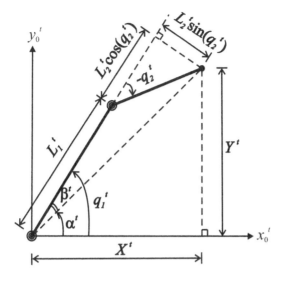

Figure 2.13 Geometric Illustration of Two-Link Thumb (Elbow up)

Accordingly, the angles α^t and β^t are gained as

$$\alpha^t = \tan^{-1}\left(\frac{Y^t}{X^t}\right),$$

$$\beta^t = -\tan^{-1}\left(\frac{L_2^t \sin(q_2^t)}{L_1^t + L_2^t \cos(q_2^t)}\right). \tag{2.3.6}$$

Then, by the summation of (2.3.6), the angle q_1^t of the joint 1 is determined as

$$q_1^t = \alpha^t + \beta^t,$$

$$= \tan^{-1}\left(\frac{Y^t}{X^t}\right) - \tan^{-1}\left(\frac{L_2^t \sin(q_2^t)}{L_1^t + L_2^t \cos(q_2^t)}\right). \tag{2.3.7}$$

Hence, inverse kinematics gives the expression of the joint angles (2.3.4 and 2.3.7).

2.3.2 Three Link Fingers

During grasping tasks, human motor control strategies and joint trajectory generation were studied by Kamper et al. [14] and Luo et al. [15]. Kamper et al. [14] showed that the fingertip trajectory generation that best fits the data recorded on the human index finger is the logarithmic form by 10 subjects (age, 21–32 years) who performed 20 grasping trails. Zollo et al. [16] showed that the polar coordinate (r, θ) is indicated as

$$r = \left[1.3394\left(L_1^i + L_2^i + L_3^i\right) - 23.255\right] \exp\left(-0.062\theta\right). \tag{2.3.8}$$

Index finger, middle finger, ring finger, and little finger are considered as three-link fingers (Section 2.1), so the inverse kinematic model of index finger is represented all three-link fingers. Based on the result of forward kinematics in Section 2.2.4, the logarithmic form expressed in the fingertip *Cartesian* coordinates (X^i, Y^i) of the index finger (i) is described in terms of three joint variables q_1^i, q_2^i, and q_3^i as

$$X^i = d + L_1^i \cos(q_1^i) + L_2^i \cos(q_1^i + q_2^i) + L_3^i \cos(q_1^i + q_2^i + q_3^i),$$
$$Y^i = L_1^i \sin(q_1^i) + L_2^i \sin(q_1^i + q_2^i) + L_3^i \sin(q_1^i + q_2^i + q_3^i). \qquad (2.3.9)$$

Here, based on practical data, the relation $q_3^i = 0.7q_2^i$ is used [14–16] to solve redundancy in the plane X^G-Y^G shown in Figure 2.10. Using this relation $q_3^i = 0.7q_2^i$ into (2.3.9), the resulting equations are expressed as

$$X^i = d + L_1^i \cos(q_1^i) + L_2^i \cos(q_1^i + q_2^i) + L_3^i \cos(q_1^i + 1.7q_2^i),$$
$$Y^i = L_1^i \sin(q_1^i) + L_2^i \sin(q_1^i + q_2^i) + L_3^i \sin(q_1^i + 1.7q_2^i). \qquad (2.3.10)$$

The resulting nonlinear functions (2.3.10) are solved by numerical methods, such as Newton–Raphson [9], the HPSONN by Wen et al. [17], GA by Chen et al. [12, 18, 19], and ANFIS by Chen et al. [12, 18, 19]. Now, setting up the two functions $f_1^i\left(q_1^i, q_2^i\right)$ and $f_2^i\left(q_1^i, q_2^i\right)$ of the two variables q_1^i and q_2^i, we get

$$f_1^i\left(q_1^i, q_2^i\right) = d + L_1^i \cos(q_1^i) + L_2^i \cos(q_1^i + q_2^i) + L_3^i \cos(q_1^i + 1.7q_2^i) - X^i,$$
$$f_2^i\left(q_1^i, q_2^i\right) = L_1^i \sin(q_1^i) + L_2^i \sin(q_1^i + q_2^i) + L_3^i \sin(q_1^i + 1.7q_2^i) - Y^i.$$

$$(2.3.11)$$

This becomes an optimal (minimization) problem with two objective functions $f_1^i(q_1^i, q_2^i)$ and $f_2^i\left(q_1^i, q_2^i\right)$ with two variables q_1^i and q_2^i. The optimal variables q_1^{i*} and q_2^{i*} are searched to make the two objective functions $f_1^i\left(q_1^i, q_2^i\right)$ and $f_2^i\left(q_1^i, q_2^i\right)$ close to zero using genetic algorithm, then q_1^{i*}, q_2^{i*}, and $q_3^{i*}(= 0.7q_2^{i*})$ are the solutions of the joint angles of the fingertip coordinate (X^i, Y^i). Alternatively, the inverse kinematics problem is solved using ANFIS method [20], where the input of the fuzzy-neuro system is the Cartesian space and the output is the joint space. During our simulations [12, 13, 18, 19], we found that the GA method although gives a better solution (error $\approx 10^{-7}$), takes more execution time whereas the ANFIS gave a good solution (error $\approx 10^{-4}$) with less time compared to GA method. ANFIS and GA will be detailed in Sections 4.3 and 4.5, respectively. Once we found the angular positions as above, then, all joint angular velocities and angular accelerations of the index finger are calculated by differential kinematics in Section 2.4. Similarly, all joint angular velocities and angular accelerations of middle finger, ring finger, and little finger are also computed.

2.3.3 Fingertip Workspace

2.3.3.1 Two Link Thumb and Three Link Index Finger When the two-link thumb and three-link fingers are doing extension/flexion motions, the ranges that all fingertips can achieve are restricted to the joint angles of MCP, PIP, and DCP. In other words, when all joints of fingers perform all possible motions, the region which is achieved by the fingertips is called *reachable workspace*. The maximum joint angles of each joint depend on different individuals. Referring to inverse kinematics, Figure 2.14 shows the workspace of the two-link thumb. The first and second

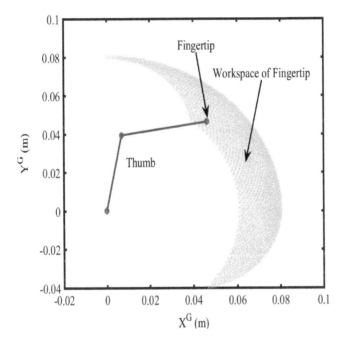

Figure 2.14 Workspace of Two-Link Thumb

joint angular positions (or joint angles) are constrained in the ranges of [0,90] and [−80,0] (deg), respectively [21]. The * region represents the workspace of the two-link thumb (we do not consider the orientation so far).

Similarly, the workspace of the three-link index finger is shown in Figure 2.15. The first (MCP), second (PIP), and third (DIP) joint angles of the index finger are constrained in the ranges of [0,90], [0,110] and [0,80] (deg), respectively [21], and the fingertip of index finger is only reachable in the * area. In addition, Figure 2.16 combines Figures 2.14 and 2.15 with a square object [12]. Thus, the lower and upper regions, respectively, show the reachable fingertip positions of thumb and index finger. Further, the overlap region represents the space both the thumb and the index finger can reach. Both the two-link lengths of thumb are given as 0.040 (m) and the three-link lengths of index finger are selected as 0.040, 0.040, and 0.030 (m) [22].

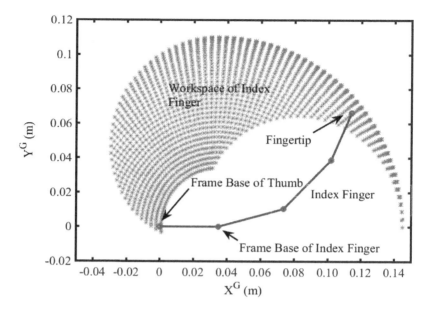

Figure 2.15   Workspace of Three-Link Index Finger

2.3.3.2 Five Fingered Robotic Hand   Figure 2.17 shows the workspace of 14-DOF, five-fingered robotic hand with a rectangular rod. The first, second, and third joint angles of the other three fingers (middle finger, ring finger, and little finger) are constrained in the ranges of [0,90], [0,110], and [0,80] (deg), respectively [21]. Thus, the gray regions, respectively, show the reachable fingertip positions of thumb, index finger, middle finger, ring finger, and little finger. Note that the X^G axis is of different scale to show more clearly the individual three-dimensional regions of reach [18].

2.4   Differential Kinematics

Sections 2.2 (forward kinematics) and 2.3 (inverse kinematics) derive the relationship between the positions of fingertips and the angular positions of joints. This section will introduce differential kinematics, which establishes the relationship between the linear and angular velocities and accelerations of fingertips and the angular velocities and accelerations of joints by the manipulator geometric Jacobian. Before calculating the geometric Jacobian, some properties of rotation matrices and rigid body kinematics need to be reviewed. Then, the geometric Jacobian of a serial n-link revolute-joint planar manipulator, two-link thumb, and three-link index finger will be computed in the following sections.

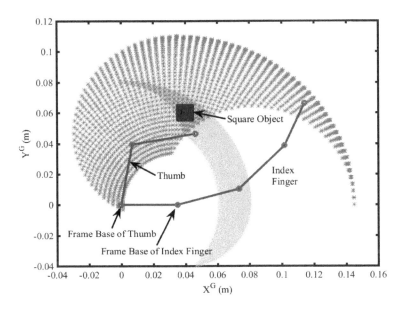

Figure 2.16   Workspace of Thumb and Index Finger with a Square Object

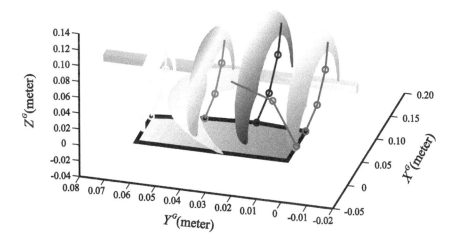

Figure 2.17   Workspace of the Five-Fingered Robotic Hand with a Rectangular Rod

2.4.1   Serial n Link Revolute Joint Planar Manipulator

According to the forward kinematics in Section 2.2, (2.2.18) is the forward kinematics equation of a serial n-link revolute-joint planar manipulator.

$$
\begin{aligned}
{}^{0}\mathbf{T}_n(\mathbf{q}) &= {}^{0}\mathbf{T}_1(\mathbf{q}) \, {}^{1}\mathbf{T}_2(\mathbf{q}) \, \cdots \, {}^{i-1}\mathbf{T}_i(\mathbf{q}) \, \cdots \, {}^{n-1}\mathbf{T}_n(\mathbf{q}), \\[2mm]
&= \begin{bmatrix}
C_{12\ldots n} & -S_{12\ldots n} & 0 & \displaystyle\sum_{i=1}^{n} L_i C_{12\ldots i} \\[4mm]
S_{12\ldots n} & C_{12\ldots n} & 0 & \displaystyle\sum_{i=1}^{n} L_i S_{12\ldots i} \\[4mm]
0 & 0 & 1 & 0 \\[1mm]
0 & 0 & 0 & 1
\end{bmatrix}, \\[2mm]
&= \begin{bmatrix}
{}^{0}\mathbf{R}_n(\mathbf{q}) & {}^{0}\mathbf{P}_n(\mathbf{q}) \\
0 & 1
\end{bmatrix},
\end{aligned}
\tag{2.4.1}
$$

where $\mathbf{q} = [q_1 \ q_2 \ \cdots \ q_n]'$ is the angular position vector of n joints; ${}^{0}\mathbf{T}_n(\mathbf{q})$ is the homogeneous transformation matrix to transfer the end-effector to frame base; ${}^{0}\mathbf{R}_n(\mathbf{q})$ and ${}^{0}\mathbf{P}_n(\mathbf{q})$ are the rotation matrix (orientation) and translation vector (position) to transfer the end-effector to frame base, respectively. The notations $C_{12\ldots n} = \cos(q_1 + q_2 + \cdots + q_n)$ and $S_{12\ldots n} = \sin(q_1 + q_2 + \cdots + q_n)$ are used.

The *linear* velocity ${}^{0}\mathbf{P}_n$ and *angular* velocity ${}^{0}\boldsymbol{\omega}_n$ of the origin of the end-effector frame are linear relations to the joint angular velocities $\dot{\mathbf{q}}$. The *linear* relation represents the differential kinematics equation and is expressed as

$$
{}^{0}\mathbf{V}_n = \mathbf{J}(\mathbf{q}) \, \dot{\mathbf{q}}.
\tag{2.4.2}
$$

Here, ${}^{0}\mathbf{V}_n = [{}^{0}\dot{\mathbf{P}}_n \ {}^{0}\boldsymbol{\omega}_n]'$ is the (6×1) end-effector velocity vector. $\mathbf{J}(\mathbf{q}) = [\mathbf{J}_{\mathbf{P}}(\mathbf{q}) \ \mathbf{J}_{\mathbf{O}}(\mathbf{q})]'$ is the $(6 \times n)$ geometric Jacobian matrix of a serial n-link revolute-joint planar manipulator; both position Jacobian $\mathbf{J}_{\mathbf{P}}(\mathbf{q})$ and orientation Jacobian $\mathbf{J}_{\mathbf{O}}(\mathbf{q})$ are the $(3 \times n)$ matrices which contribute the joint angular velocities $\dot{\mathbf{q}}$ to the linear velocity ${}^{0}\dot{\mathbf{P}}_n$ and angular velocity ${}^{0}\boldsymbol{\omega}_n$ of the end-effector, respectively. Before calculating the geometric Jacobian $\mathbf{J}(\mathbf{q})$, some properties of rotation matrices and rigid body kinematics need be reviewed.

2.4.1.1 Some Properties of Rotation Matrices

(2.4.1) explains that the end-effector pose is a function of the joint angular position vector $\mathbf{q}$ in terms of the translation vector ${}^{0}\mathbf{P}_n(\mathbf{q})$ (position) and the rotation matrix ${}^{0}\mathbf{R}_n(\mathbf{q})$ (orientation). The goal of differential kinematics is to characterize the linear velocity ${}^{0}\dot{\mathbf{P}}_n$ and angular velocity ${}^{0}\boldsymbol{\omega}_n$ of the end-effector, so it is necessary to consider the derivative properties of the time-dependent rotation matrices with respect to time.

A time-dependent rotation matrix $\mathbf{R}(t)$ is an orthogonal matrix, so one has the property

$$
\mathbf{R}(t) \, \mathbf{R}'(t) = \mathbf{I}.
\tag{2.4.3}
$$

Differentiate (2.4.3) with respect to time and obtain the equation

$$
\dot{\mathbf{R}}(t) \, \mathbf{R}'(t) + \mathbf{R}(t) \, \dot{\mathbf{R}}'(t) = 0.
\tag{2.4.4}
$$

Set $S(t) = \dot{R}(t) \; R'(t)$ and (2.4.4) is rewritten as

$$S(t) + S'(t) \;=\; 0. \tag{2.4.5}$$

Therefore, $S(t)$ is the (3×3) skew-symmetric matrix. Postmultiply $R(t)$ on both sides of (2.4.4) to give

$$\dot{R}(t) \; R'(t) \; R(t) \;=\; -R(t) \; \dot{R}'(t) \; R(t). \tag{2.4.6}$$

Take (2.4.3) and (2.4.5) into the left and right sides of (2.4.6) to give

$$\begin{aligned} \dot{R}(t) \;&=\; -(R(t) \; \dot{R}'(t)) \; R(t), \\ &=\; S(t) \; R(t), \end{aligned} \tag{2.4.7}$$

which gives the first derivative of $R(t)$ with respect to time. Consider the vector $p(t) = R(t) \; p'$, where p' is a constant vector. Then, the first derivative of $R(t)$ with respect to time is expressed as

$$\dot{p}(t) = \dot{R}(t) \; p'. \tag{2.4.8}$$

Taking (2.4.7) into (2.4.8), (2.4.8) is rewritten as

$$\dot{p}(t) = S(t) \; R(t) \; p'. \tag{2.4.9}$$

If the vector $\omega(t)$ represents the angular velocity of frame $R(t)$ with respect to the reference frame at time t, then one has the relation

$$\dot{p}(t) = \omega(t) \times R(t) \; p'. \tag{2.4.10}$$

Hence, comparing (2.4.9) and (2.4.10), we know $S(t)$ is the vector product between $\omega(t) = [\omega_x \;\; \omega_y \;\; \omega_z]'$ and $R(t) \; p'$. $S(t)$ is expressed as

$$\begin{aligned} S(t) \;&=\; \begin{bmatrix} 0 & -\omega_z & \omega_y \\ \omega_z & 0 & -\omega_x \\ -\omega_y & \omega_x & 0 \end{bmatrix}, \\ &=\; S(\omega(t)). \end{aligned} \tag{2.4.11}$$

Thus, (2.4.7) is rewritten as

$$\dot{R}(t) \;=\; S(\omega(t)) \; R(t). \tag{2.4.12}$$

Besides, the equation below is shown by [10].

$$R(t) \; S(\omega(t)) \; R'(t) \;=\; S(R(t) \; \omega(t)). \tag{2.4.13}$$

2.4.1.2 Rigid Body Kinematics   As shown in Figure 2.3, (2.2.6) gives the coordinate transformation of translation vector $^0o_1$ and rotation matrix $^0R_1$ from frame 1 to frame 0.

$$\mathbf{p}^0 = {}^0\mathbf{o}_1 + {}^0\mathbf{R}_1\ \mathbf{p}^1. \tag{2.4.14}$$

Take the first derivative of (2.4.14) with respect to time and obtain

$$\dot{\mathbf{p}}^0 = {}^0\dot{\mathbf{o}}_1 + {}^0\mathbf{R}_1\ \dot{\mathbf{p}}^1 + {}^0\dot{\mathbf{R}}_1\ \mathbf{p}^1. \tag{2.4.15}$$

Substituting (2.4.12) into (2.4.15), (2.4.15) is rewritten as

$$\begin{aligned}
\dot{\mathbf{p}}^0 &= {}^0\dot{\mathbf{o}}_1 + {}^0\mathbf{R}_1\ \dot{\mathbf{p}}^1 + \mathbf{S}({}^0\boldsymbol{\omega}_1)\ {}^0\mathbf{R}_1\ \mathbf{p}^1, \\
&= {}^0\dot{\mathbf{o}}_1 + {}^0\mathbf{R}_1\ \dot{\mathbf{p}}^1 + {}^0\boldsymbol{\omega}_1 \times {}^0\mathbf{p}_1,
\end{aligned} \tag{2.4.16}$$

where $^0\mathbf{p}_1 = {}^0\mathbf{R}_1\ \mathbf{p}^1$.

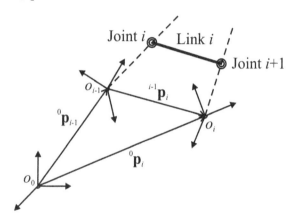

Figure 2.18   Vector Representation of Link i

Figure 2.18 shows the vector representation of link i, which connects joints i and $i + 1$. Frame $i - 1$ is attached to link $i - 1$ and has the origin o_{i-1} along joint i while frame i is attached to link i and has the origin o_i along joint $i + 1$. $^0\mathbf{p}_{i-1}$ and $^0\mathbf{p}_i$ represent the position vectors of frames $i - 1$ and i with respect to frame base, respectively. According to (2.4.14), the coordinate transformation transferring from frame $i - 1$ to frame i is denoted as

$$^0\mathbf{p}_i = {}^0\mathbf{p}_{i-1} + {}^0\mathbf{R}_{i-1}\ {}^{i-1}\mathbf{p}_i^{i-1}, \tag{2.4.17}$$

where $^{i-1}\mathbf{p}_i^{i-1}$ is the position vector of the frame origin o_i (right subscript) with respect to the frame origin o_{i-1} (left superscript) expressed in frame $i - 1$ (right superscript). Differentiating (2.4.17) with respect to time gives

$$\begin{aligned}
^0\dot{\mathbf{p}}_i &= {}^0\dot{\mathbf{p}}_{i-1} + {}^0\mathbf{R}_{i-1}\ {}^{i-1}\dot{\mathbf{p}}_i^{i-1} + {}^0\dot{\mathbf{R}}_{i-1}\ {}^{i-1}\mathbf{p}_i^{i-1}, \\
&= {}^0\dot{\mathbf{p}}_{i-1} + {}^{i-1}\dot{\mathbf{p}}_i + {}^0\boldsymbol{\omega}_{i-1} \times {}^{i-1}\mathbf{p}_i,
\end{aligned} \tag{2.4.18}$$

which expresses the linear velocity $^0\dot{\mathbf{p}}_i$ of link i as a function of the translational velocity $(^0\dot{\mathbf{p}}_{i-1} + {}^{i-1}\dot{\mathbf{p}}_i)$ and rotational velocity $(^0\boldsymbol{\omega}_{i-1} \times {}^{i-1}\mathbf{p}_i)$ of link $i-1$. Next, the link angular velocity will be derived. The rotation composition is

$$^0\mathbf{R}_i = {}^0\mathbf{R}_{i-1}\,{}^{i-1}\mathbf{R}_i. \qquad (2.4.19)$$

Differentiating (2.4.19) with respect to time gives

$$^0\dot{\mathbf{R}}_i = {}^0\dot{\mathbf{R}}_{i-1}\,{}^{i-1}\mathbf{R}_i + {}^0\mathbf{R}_{i-1}\,{}^{i-1}\dot{\mathbf{R}}_i. \qquad (2.4.20)$$

Substituting (2.4.12) into (2.4.20) gives

$$\begin{aligned}
\mathbf{S}(^0\boldsymbol{\omega}_i)\,{}^0\mathbf{R}_i &= \mathbf{S}(^0\boldsymbol{\omega}_{i-1})\,{}^0\mathbf{R}_{i-1}\,{}^{i-1}\mathbf{R}_i + {}^0\mathbf{R}_{i-1}\,\mathbf{S}(^{i-1}\boldsymbol{\omega}_i^{i-1})\,{}^{i-1}\mathbf{R}_i, \\
&= \mathbf{S}(^0\boldsymbol{\omega}_{i-1})\,{}^0\mathbf{R}_{i-1}\,{}^{i-1}\mathbf{R}_i + {}^0\mathbf{R}_{i-1}\,\mathbf{S}(^{i-1}\boldsymbol{\omega}_i^{i-1})\,{}^0\mathbf{R'}_{i-1}\,{}^0\mathbf{R}_{i-1}\,{}^{i-1}\mathbf{R}_i, \\
&= \mathbf{S}(^0\boldsymbol{\omega}_{i-1})\,{}^0\mathbf{R}_i + \mathbf{S}(^0\mathbf{R}_{i-1}\,{}^{i-1}\boldsymbol{\omega}_i^{i-1})\,{}^0\mathbf{R}_i, \\
&= \mathbf{S}(^0\boldsymbol{\omega}_{i-1})\,{}^0\mathbf{R}_i + \mathbf{S}(^{i-1}\boldsymbol{\omega}_i)\,{}^0\mathbf{R}_i, \qquad (2.4.21)
\end{aligned}$$

which leads to the result

$$^0\boldsymbol{\omega}_i = {}^0\boldsymbol{\omega}_{i-1} + {}^{i-1}\boldsymbol{\omega}_i, \qquad (2.4.22)$$

which expresses the angular velocity $^0\boldsymbol{\omega}_i$ of link i as a function of the angular velocity $^0\boldsymbol{\omega}_{i-1}$ of link $i-1$ and the angular velocity $^{i-1}\boldsymbol{\omega}_i$ of link i with respect to link $i-1$.

For revolute joint, the rotation of frame i with respect to frame $i-1$ is induced by the motion of joint i. Therefore, the second term on the right side of (2.4.18) can be written as

$$^{i-1}\dot{\mathbf{p}}_i = {}^{i-1}\boldsymbol{\omega}_i \times {}^{i-1}\mathbf{p}_i. \qquad (2.4.23)$$

Substituting (2.4.23) into (2.4.18) gives

$$\begin{aligned}
^0\dot{\mathbf{p}}_i &= {}^0\dot{\mathbf{p}}_{i-1} + {}^{i-1}\boldsymbol{\omega}_i \times {}^{i-1}\mathbf{p}_i + {}^0\boldsymbol{\omega}_{i-1} \times {}^{i-1}\mathbf{p}_i, \\
&= {}^0\dot{\mathbf{p}}_{i-1} + (^0\boldsymbol{\omega}_{i-1} + {}^{i-1}\boldsymbol{\omega}_i) \times {}^{i-1}\mathbf{p}_i, \\
&= {}^0\dot{\mathbf{p}}_{i-1} + {}^0\boldsymbol{\omega}_i \times {}^{i-1}\mathbf{p}_i. \qquad (2.4.24)
\end{aligned}$$

(2.4.22) is rewritten as

$$^0\boldsymbol{\omega}_i = {}^0\boldsymbol{\omega}_{i-1} + \dot{q}_i\,\vec{z}_{i-1}, \qquad (2.4.25)$$

where $\vec{z}_{i-1}$ is the unit vector of joint i axis.

Substituting (2.4.24) and (2.4.25) into (2.4.2) gives

$$
\begin{bmatrix} {}^0\dot{\mathbf{P}}_n \\ {}^0\boldsymbol{\omega}_n \end{bmatrix} = \begin{bmatrix} \mathbf{J}_\mathbf{P}(\mathbf{q}) \\ \mathbf{J}_\mathbf{O}(\mathbf{q}) \end{bmatrix} \dot{\mathbf{q}} = \begin{bmatrix} \mathbf{J}_{P1} & \mathbf{J}_{P2} & \cdots & \mathbf{J}_{Pn} \\ \mathbf{J}_{O1} & \mathbf{J}_{O2} & \cdots & \mathbf{J}_{On} \end{bmatrix} \begin{bmatrix} \dot{q}_1 \\ \dot{q}_2 \\ \vdots \\ \dot{q}_n \end{bmatrix},
$$

$$
= \begin{bmatrix} \displaystyle\sum_{i=1}^n \mathbf{J}_{Pi}\dot{q}_i \\ \displaystyle\sum_{i=1}^n \mathbf{J}_{Oi}\dot{q}_i \end{bmatrix} = \begin{bmatrix} \displaystyle\sum_{i=1}^n \dfrac{\partial\,{}^0\mathbf{P}_n}{\partial\,q_i}\dot{q}_i \\ \displaystyle\sum_{i=1}^n {}^{i-1}\boldsymbol{\omega}_i \end{bmatrix} = \begin{bmatrix} \displaystyle\sum_{i=1}^n {}^{i-1}\boldsymbol{\omega}_i \times {}^{i-1}\mathbf{p}_n \\ \displaystyle\sum_{i=1}^n \dot{q}_i \vec{z}_{i-1} \end{bmatrix},
$$

$$
= \begin{bmatrix} \displaystyle\sum_{i=1}^n \dot{q}_i \vec{z}_{i-1} \times ({}^0\mathbf{P}_n - {}^0\mathbf{p}_{i-1}) \\ \displaystyle\sum_{i=1}^n \dot{q}_i \vec{z}_{i-1} \end{bmatrix}. \tag{2.4.26}
$$

Therefore, the geometric Jacobian matrix $\mathbf{J}(\mathbf{q})$ is denoted as

$$
\mathbf{J}(\mathbf{q}) = \begin{bmatrix} \mathbf{J}_\mathbf{P}(\mathbf{q}) \\ \mathbf{J}_\mathbf{O}(\mathbf{q}) \end{bmatrix} = \begin{bmatrix} \mathbf{J}_{P1} & \mathbf{J}_{P2} & \cdots & \mathbf{J}_{Pn} \\ \mathbf{J}_{O1} & \mathbf{J}_{O2} & \cdots & \mathbf{J}_{On} \end{bmatrix}, \tag{2.4.27}
$$

where $[\mathbf{J}_{Pi}\ \mathbf{J}_{Oi}]'$ $(i = 1, 2, \cdots, n)$ is computed by

$$
\begin{bmatrix} \mathbf{J}_{Pi} \\ \mathbf{J}_{Oi} \end{bmatrix} = \begin{bmatrix} \vec{z}_{i-1} \times ({}^0\mathbf{P}_n - {}^0\mathbf{p}_{i-1}) \\ \vec{z}_{i-1} \end{bmatrix}. \tag{2.4.28}
$$

The unit vector $\vec{z}_{i-1}$ of joint i axis is obtained from the third column of the rotation matrix ${}^0\mathbf{R}_{i-1}$ and is expressed as

$$
\vec{z}_{i-1} = {}^0\mathbf{R}_1(q_1)\,{}^1\mathbf{R}_2(q_2) \cdots {}^{i-2}\mathbf{R}_{i-1}(q_{i-1})\,\vec{z}_0, \tag{2.4.29}
$$

where $\vec{z}_0 = [0\ \ 0\ \ 1]' = \vec{z}_j$ $(j = 1, 2, \cdots, n)$.

The position vector ${}^0\mathbf{P}_n$ of the end-effector is obtained from the first three entries of the fourth column (translation vector) of the homogeneous transformation matrix ${}^0\mathbf{T}_n$ and is expressed as

$$
\begin{aligned}
{}^0\widetilde{\mathbf{P}}_n &= {}^0\mathbf{T}_n\,{}^0\widetilde{\mathbf{p}}_0, \\
&= {}^0\mathbf{T}_1(q_1)\,{}^1\mathbf{T}_2(q_2) \cdots {}^{n-1}\mathbf{T}_n(q_n)\,{}^0\widetilde{\mathbf{p}}_0, \tag{2.4.30}
\end{aligned}
$$

where ${}^0\widetilde{\mathbf{P}}_n = [{}^0\mathbf{P}_n\ \ 1]'$ and ${}^0\widetilde{\mathbf{p}}_0 = [0\ \ 0\ \ 0\ \ 1]'$. Hence, ${}^0\mathbf{P}_n$ is calculated by

$$
{}^0\mathbf{P}_n = \begin{bmatrix} \displaystyle\sum_{i=1}^n L_i C_{12\ldots i} \\ \displaystyle\sum_{i=1}^n L_i S_{12\ldots i} \\ 0 \end{bmatrix}. \tag{2.4.31}
$$

Here, $C_{12...i} = \cos(q_1 + q_2 + \cdots + q_i)$ and $S_{12...i} = \sin(q_1 + q_2 + \cdots + q_i)$.
Similarly, the position vector $^0\mathbf{p}_{i-1}$ of the joint i is obtained from the first three entries of the fourth column (translation vector) of the homogeneous transformation matrix $^0\mathbf{T}_{i-1}$ and is expressed as

$$
\begin{aligned}
^0\widetilde{\mathbf{p}}_{i-1} &= {}^0\mathbf{T}_{i-1} \,{}^0\widetilde{\mathbf{p}}_0, \\
&= {}^0\mathbf{T}_1(q_1) \,{}^1\mathbf{T}_2(q_2) \cdots {}^{i-2}\mathbf{T}_{i-1}(q_{i-1}) \,{}^0\widetilde{\mathbf{p}}_0,
\end{aligned}
\tag{2.4.32}
$$

where $^0\widetilde{\mathbf{p}}_{i-1} = [\,{}^0\mathbf{p}_{i-1} \ \ 1]'$. Thus, $^0\mathbf{p}_{i-1}$ $(i = 2, 3, \cdots, n)$ is computed by

$$
^0\mathbf{p}_{i-1} = \begin{bmatrix} \displaystyle\sum_{j=1}^{i-1} L_j C_{12...(i-1)} \\[2ex] \displaystyle\sum_{j=1}^{i-1} L_j S_{12...(i-1)} \\[2ex] 0 \end{bmatrix}.
\tag{2.4.33}
$$

Here, $C_{12...(i-1)} = \cos(q_1 + q_2 + \cdots + q_{i-1})$ and $S_{12...(i-1)} = \sin(q_1 + q_2 + \cdots + q_{i-1})$. Substituting (2.4.29), (2.4.31), and (2.4.33) into (2.4.28), the geometric Jacobian $\mathbf{J}(\mathbf{q})$ in (2.4.27) is rewritten as

$$
\begin{aligned}
\mathbf{J}(\mathbf{q}) &= \begin{bmatrix} \mathbf{J}_{P1} & \mathbf{J}_{P2} & \cdots & \mathbf{J}_{Pn} \\ \mathbf{J}_{O1} & \mathbf{J}_{O2} & \cdots & \mathbf{J}_{On} \end{bmatrix}, \\[2ex]
&= \begin{bmatrix} \vec{z}_0 \times ({}^0\mathbf{P}_n - {}^0\mathbf{p}_0) & \vec{z}_1 \times ({}^0\mathbf{P}_n - {}^0\mathbf{p}_1) & \cdots & \vec{z}_n \times ({}^0\mathbf{P}_n - {}^0\mathbf{p}_{n-1}) \\ \vec{z}_0 & \vec{z}_1 & \cdots & \vec{z}_n \end{bmatrix}, \\[2ex]
&= \begin{bmatrix} -\displaystyle\sum_{i=1}^{n} L_i S_{12...n} & -\displaystyle\sum_{i=2}^{n} L_i S_{12...n} & \cdots & -\displaystyle\sum_{i=n}^{n} L_i S_{12...n} \\[2ex] \displaystyle\sum_{i=1}^{n} L_i C_{12...n} & \displaystyle\sum_{i=2}^{n} L_i C_{12...n} & \cdots & \displaystyle\sum_{i=n}^{n} L_i C_{12...n} \\[1ex] 0 & 0 & \cdots & 0 \\ 0 & 0 & \cdots & 0 \\ 0 & 0 & \cdots & 0 \\ 1 & 1 & \cdots & 1 \end{bmatrix}.
\end{aligned}
\tag{2.4.34}
$$

The entry J_{ij} $(i, j \in N)$ of the geometric Jacobian $\mathbf{J}(\mathbf{q})$ in (2.4.2) is rewritten as

$$
J_{ij} = \begin{cases} -\displaystyle\sum_{j=1}^{n} L_j \sin\left(\displaystyle\sum_{k=1}^{j} q_k\right) & \text{if } i = 1, j \in [1, 6]; \\[3ex] \displaystyle\sum_{j=1}^{n} L_j \cos\left(\displaystyle\sum_{k=1}^{j} q_k\right) & \text{if } i = 2, j \in [1, 6]; \\[3ex] 0 & \text{if } i \in [3, 5], j \in [1, 6]; \\[1ex] 1 & \text{if } i = 6, j \in [1, 6]. \end{cases}
$$

Then, the angular velocities $\dot{\mathbf{q}}$ are written as

$$\dot{\mathbf{q}} = \mathbf{J}^{-1}(\mathbf{q}) \; {}^0\mathbf{V}_n. \qquad (2.4.35)$$

Similarly, the angular accelerations $\ddot{\mathbf{q}}$ are obtained as

$$\ddot{\mathbf{q}} = \mathbf{J}^{-1}(\mathbf{q}) \left({}^0\mathbf{A}_n - \dot{\mathbf{J}}(\mathbf{q})\dot{\mathbf{q}} \right). \qquad (2.4.36)$$

Here, ${}^0\mathbf{A}_n = [{}^0\ddot{\mathbf{P}}_n \quad {}^0\boldsymbol{\alpha}_n]'$ is the (6×1) end-effector acceleration vector, including linear acceleration ${}^0\ddot{\mathbf{P}}_n$ and angular acceleration ${}^0\boldsymbol{\alpha}_n$. The differential Jacobian $\dot{\mathbf{J}}(\mathbf{q})$ is also written as [5]

$$\dot{J}_{ij} = \begin{cases} -\displaystyle\sum_{j=1}^{n} L_j \cos\left(\displaystyle\sum_{k=1}^{j} q_k\right)\left(\displaystyle\sum_{k=1}^{j} \dot{q}_k\right) & \text{if } i = 1, j \in [1,6]; \\[2ex] -\displaystyle\sum_{j=1}^{n} L_j \sin\left(\displaystyle\sum_{k=1}^{j} q_k\right)\left(\displaystyle\sum_{k=1}^{j} \dot{q}_k\right) & \text{if } i = 1, j \in [1,6]; \\[2ex] 0 & \text{if } i \in [3,6], j \in [1,6]. \end{cases}$$

2.4.2 Two Link Thumb

If we only consider the linear velocities of the fingertip, then the angular velocities and angular accelerations of joints are obtained by taking the first derivative and second derivative on (2.3.1). The corresponding *linear* velocities $\dot{\mathbf{P}}^t$ ($\dot{X}^t$, $\dot{Y}^t$) = $d(X^t, Y^t)/dt$ of the fingertip are obtained as

$$\begin{bmatrix} \dot{X}^t \\ \dot{Y}^t \end{bmatrix} = \begin{bmatrix} -L_1^t \sin(q_1^t) - L_2^t \sin(q_1^t + q_2^t) & -L_2^t \sin(q_1^t + q_2^t) \\ L_1^t \cos(q_1^t) + L_2^t \cos(q_1^t + q_2^t) & L_2^t \cos(q_1^t + q_2^t) \end{bmatrix} \begin{bmatrix} \dot{q}_1^t \\ \dot{q}_2^t \end{bmatrix},$$

or the matrix form

$$\dot{\mathbf{P}}^t = \mathbf{J}_P^t(\mathbf{q}^t) \, \dot{\mathbf{q}}^t. \qquad (2.4.37)$$

The matrices $\dot{\mathbf{P}}^t$, $\dot{\mathbf{q}}^t$, $\mathbf{q}^t$, and $\mathbf{J}_P^t(\mathbf{q}^t)$ are

$$\dot{\mathbf{P}}^t = \begin{bmatrix} \dot{X}^t \\ \dot{Y}^t \end{bmatrix}, \quad \dot{\mathbf{q}}^t = \begin{bmatrix} \dot{q}_1^t \\ \dot{q}_2^t \end{bmatrix}, \quad \mathbf{q}^t = \begin{bmatrix} q_1^t \\ q_2^t \end{bmatrix},$$

$$\mathbf{J}_P^t(\mathbf{q}^t) = \begin{bmatrix} -L_1^t \sin(q_1^t) - L_2^t \sin(q_1^t + q_2^t) & -L_2^t \sin(q_1^t + q_2^t) \\ L_1^t \cos(q_1^t) + L_2^t \cos(q_1^t + q_2^t) & L_2^t \cos(q_1^t + q_2^t) \end{bmatrix}.$$

$$(2.4.38)$$

The matrix $\mathbf{J}_P^t(\mathbf{q}^t)$ is called a submatrix of the *geometric Jacobian* of the thumb. The complete geometric Jacobian, such as (2.4.34), would be a (6×2) matrix, with the last three rows that account for the angular velocities of each link. We only consider

the geometric Jacobian for the planar location $\mathbf{P}^t$ (X^t, Y^t) and not the orientation, that is, a submatrix of the geometric Jacobian.
The angular velocities $\dot{q}_1^t$ and $\dot{q}_2^t$ of the joints 1 and 2 are

$$\dot{\mathbf{q}}^t = \mathbf{J}_P^t(\mathbf{q}^t)^{-1} \, \dot{\mathbf{P}}^t. \tag{2.4.39}$$

Similarly, the angular accelerations $\ddot{q}_1^t$ and $\ddot{q}_2^t$ of the joints 1 and 2 are obtained as

$$\ddot{\mathbf{q}}^t = \mathbf{J}_P^t(\mathbf{q}^t)^{-1} \left(\ddot{\mathbf{P}}^t - \frac{d\mathbf{J}_P^t(\mathbf{q}^t)}{dt} \dot{\mathbf{q}}^t \right), \tag{2.4.40}$$

where $\ddot{\mathbf{P}}^t$ is the *linear* acceleration vector of the fingertip. $\ddot{\mathbf{P}}^t$, $\ddot{\mathbf{q}}^t$, and $d\mathbf{J}_P^t(\mathbf{q}^t)/dt$ are denoted as

$$\ddot{\mathbf{P}}^t = \begin{bmatrix} \ddot{X}^t \\ \ddot{Y}^t \end{bmatrix}, \quad \ddot{\mathbf{q}}^t = \begin{bmatrix} \ddot{q}_1^t \\ \ddot{q}_2^t \end{bmatrix}, \tag{2.4.41}$$

$$\frac{d\mathbf{J}_P^t(\mathbf{q}^t)}{dt} = \begin{bmatrix} -L_1^t \cos(q_1^t)\dot{q}_1^t - L_2^t \cos(q_1^t + q_2^t)\left(\dot{q}_1^t + \dot{q}_2^t\right) & -L_2^t \cos(q_1^t + q_2^t)\left(\dot{q}_1^t + \dot{q}_2^t\right) \\ -L_1^t \sin(q_1^t)\dot{q}_1^t - L_2^t \sin(q_1^t + q_2^t)\left(\dot{q}_1^t + \dot{q}_2^t\right) & -L_2^t \sin(q_1^t + q_2^t)\left(\dot{q}_1^t + \dot{q}_2^t\right) \end{bmatrix} \tag{2.4.42}$$

2.4.3 Three Link Index Finger

Similar to two-link thumb, by taking the first derivative of the fingertip positions, we get the *linear* velocities $d(X^i, Y^i)/dt$, which are expressed as

$$\dot{\mathbf{P}}^i = \mathbf{J}_P^i(\mathbf{q}^i) \, \dot{\mathbf{q}}^i. \tag{2.4.43}$$

The matrices $\dot{\mathbf{P}}^i$, $\dot{\mathbf{q}}^i$, $\mathbf{q}^i$, and $\mathbf{J}^i(\mathbf{q})$ follow as

$$\dot{\mathbf{P}}^i = \begin{bmatrix} \dot{X}^i \\ \dot{Y}^i \end{bmatrix}, \quad \dot{\mathbf{q}}^i = \begin{bmatrix} \dot{q}_1^i \\ \dot{q}_2^i \end{bmatrix}, \quad \mathbf{q}^i = \begin{bmatrix} q_1^i \\ q_2^i \end{bmatrix},$$

$$\mathbf{J}^i(\mathbf{q}^i) = \begin{bmatrix} J_{11}^i(\mathbf{q}^i) & J_{12}^i(\mathbf{q}^i) \\ J_{21}^i(\mathbf{q}^i) & J_{22}^i(\mathbf{q}^i) \end{bmatrix} \tag{2.4.44}$$

with

$$
\begin{aligned}
J_{11}^i(\mathbf{q}^i) &= -L_1^i \sin(q_1^i) - L_2^i \sin(q_1^i + q_2^i) - L_3^i \sin(q_1^i + 1.7q_2^i), \\
J_{12}^i(\mathbf{q}^i) &= -L_2^i \sin(q_1^i + q_2^i) - 1.7L_3^i \sin(q_1^i + 1.7q_2^i), \\
J_{21}^i(\mathbf{q}^i) &= L_1^i \cos(q_1^i) + L_2^i \cos(q_1^i + q_2^i) + L_3^i \cos(q_1^i + 1.7q_2^i), \\
J_{22}^i(\mathbf{q}^i) &= L_2^i \cos(q_1^i + q_2^i) + 1.7L_3^i \cos(q_1^i + 1.7q_2^i).
\end{aligned}
\tag{2.4.45}
$$

Consequently, the angular velocities $\dot{q}_1^i$ and $\dot{q}_2^i$ of the joints 1 and 2 are solved as

$$\dot{\mathbf{q}}^i = \mathbf{J}_P^i(\mathbf{q}^i)^{-1}\dot{\mathbf{P}}^i. \tag{2.4.46}$$

Simultaneously, the angular velocity $\dot{q}_3^i$ of the joint 3 is obtained as

$$\dot{q}_3^i = 0.7\,\dot{q}_2^i. \tag{2.4.47}$$

In the similar manner, the angular accelerations $\ddot{q}_1^i$ and $\ddot{q}_2^i$ of the joints 1 and 2 are obtained as

$$\ddot{\mathbf{q}}^i = \mathbf{J}_P^i(\mathbf{q}^i)^{-1}\left(\ddot{\mathbf{P}}^i - \frac{d\mathbf{J}_P^i(\mathbf{q}^i)}{dt}\dot{\mathbf{q}}^i\right), \tag{2.4.48}$$

where $\ddot{\mathbf{P}}^i$ is the *linear* acceleration vector of the fingertip. $\ddot{\mathbf{P}}^i$, $\ddot{\mathbf{q}}^i$, and $d\mathbf{J}_P^i(\mathbf{q}^i)/dt$ are

$$\ddot{\mathbf{P}}^i = \begin{bmatrix} \ddot{X}^i \\ \ddot{Y}^i \end{bmatrix}, \quad \ddot{\mathbf{q}}^i = \begin{bmatrix} \ddot{q}_1^i \\ \ddot{q}_2^i \end{bmatrix}, \quad \frac{d\mathbf{J}_P^i(\mathbf{q}^i)}{dt} = \begin{bmatrix} \dot{J}_{11}^i(\mathbf{q}^i) & \dot{J}_{12}^i(\mathbf{q}^i) \\ \dot{J}_{21}^i(\mathbf{q}^i) & \dot{J}_{22}^i(\mathbf{q}^i) \end{bmatrix} \tag{2.4.49}$$

with

$$\dot{J}_{11}^i(\mathbf{q}^i) = -L_1^i\cos(q_1^i)\dot{q}_1^i - L_2^i\cos(q_1^i + q_2^i)\left(\dot{q}_1^i + \dot{q}_2^i\right) - L_3^i\cos(q_1^i + 1.7q_2^i)\left(\dot{q}_1^i + 1.7\dot{q}_2^i\right),$$
$$\dot{J}_{12}^i(\mathbf{q}^i) = -L_2^i\cos(q_1^i + q_2^i)\left(\dot{q}_1^i + \dot{q}_2^i\right) - 1.7L_3^i\cos(q_1^i + 1.7q_2^i)\left(\dot{q}_1^i + 1.7\dot{q}_2^i\right),$$
$$\dot{J}_{21}^i(\mathbf{q}^i) = -L_1^i\sin(q_1^i)\dot{q}_1^i - L_2^i\sin(q_1^i + q_2^i)\left(\dot{q}_1^i + \dot{q}_2^i\right) - L_3^i\sin(q_1^i + 1.7q_2^i)\left(\dot{q}_1^i + 1.7\dot{q}_2^i\right),$$
$$\dot{J}_{22}^i(\mathbf{q}^i) = -L_2^i\sin(q_1^i + q_2^i)\left(\dot{q}_1^i + \dot{q}_2^i\right) - 1.7L_3^i\sin(q_1^i + 1.7q_2^i)\left(\dot{q}_1^i + 1.7\dot{q}_2^i\right).$$

$$\tag{2.4.50}$$

Then, the angular acceleration $\ddot{q}_3^i$ of the joint 3 is also obtained as

$$\ddot{q}_3^i = 0.7\,\ddot{q}_2^i. \tag{2.4.51}$$

2.5 Trajectory Planning

It is necessary to know the desired paths before the robotic hand is controlled in order to execute a specific hand motion task. Trajectory planning involves the generation of the desired paths or reference inputs, so trajectory planning is associated with some situations, such as avoiding barriers and making sure that the desired path does not exceed the voltage and torque limitations of the actuators. The simplest hand motion is *point to point* motion. Sections 2.5.1 and 2.5.2, respectively, explain a time sequence of cubic polynomial and Bézier curve functions generating a desired path described by initial and terminal conditions, including positions and velocities.

2.5.1   Trajectory Planning Using Cubic Polynomial

A polynomial function is a function defined by evaluating a polynomial and a general polynomial function $P(t)$ of variable t is written as

$$P(t) \;=\; \sum_{i=0}^{n} A_i \, t^i, \tag{2.5.1}$$

where n is a non-negative integer and A_i ($i = 0, 1, 2, \cdots, n$) are constant coefficients.

To generate smooth trajectories, a cubic polynomial function for the fingertip space is created as shown in Figure 2.19. A time history of desired (d) fingertip

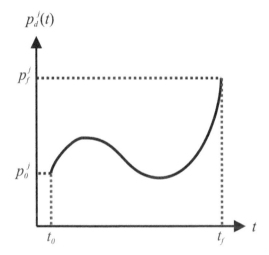

Figure 2.19   Fingertip Space Trajectory

positions (p), velocities (v), and accelerations (a) is given as [9, 10, 23]

$$
\begin{aligned}
p_d^j(t) &= A_0 + A_1 t + A_2 t^2 + A_3 t^3, \\
v_d^j(t) &= A_1 + 2A_2 t + 3A_3 t^2, \\
a_d^j(t) &= 2A_2 + 6A_3 t.
\end{aligned}
\tag{2.5.2}
$$

Here, A_0–A_3 are undetermined constants and the superscript j indicates the index of each finger, for example, $j = t, i, m, r$, and l represent thumb, index finger, middle finger, ring finger, and little finger, respectively.

Suppose we need to satisfy the following four constraint conditions at t_0 and t_f

$$
\begin{aligned}
p_d^j(t_0) &= p_0^j, \\
v_d^j(t_0) &= v_0^j, \\
p_d^j(t_f) &= p_f^j, \\
v_d^j(t_f) &= v_f^j,
\end{aligned}
\tag{2.5.3}
$$

where t_0 and t_f are the initial time and terminal time, respectively; p_0^j and p_f^j are the initial and terminal fingertip positions of the finger j. Based on the four constraint conditions at t_0 and t_f, (2.5.2), they are written as

$$
\begin{aligned}
p_d^j(t_0) &= A_0 + A_1 t_0 + A_2 t_0^2 + A_3 t_0^3 = p_0^j, \\
v_d^j(t_0) &= A_1 + 2A_2 t_0 + 3A_3 t_0^2 = v_0^j, \\
p_d^j(t_f) &= A_0 + A_1 t_f + A_2 t_f^2 + A_3 t_f^3 = p_f^j, \\
v_d^j(t_f) &= A_1 + 2A_2 t_f + 3A_3 t_f^2 = v_f^j.
\end{aligned}
\tag{2.5.4}
$$

In the matrix form, the relations (2.5.4) are rewritten as

$$
\mathbf{T}^P \, \mathbf{A} \;=\; \mathbf{P},
\tag{2.5.5}
$$

where the matrices $\mathbf{T}^P$, $\mathbf{A}$, and $\mathbf{P}$ are

$$
\mathbf{T}^P \;=\;
\begin{bmatrix}
1 & t_0 & t_0^2 & t_0^3 \\
0 & 1 & 2t_0 & 3t_0^2 \\
1 & t_f & t_f^2 & t_f^3 \\
0 & 1 & 2t_f & 3t_f^2
\end{bmatrix},
$$

$$
\mathbf{A} \;=\; \begin{bmatrix} A_0 & A_1 & A_2 & A_3 \end{bmatrix}',
$$

$$
\mathbf{P} \;=\; \begin{bmatrix} p_0^j & v_0^j & p_f^j & v_f^j \end{bmatrix}'.
\tag{2.5.6}
$$

Therefore, the four unknown constants, A_0–A_3, are computed by

$$
\mathbf{A} \;=\; \mathbf{T}^{P-1} \, \mathbf{P}.
\tag{2.5.7}
$$

2.5.2   Trajectory Planning Using Cubic Bézier Curve

Bézier curves are widely used in computer graphics to model smooth curves. Generalizations of Bézier curves to higher dimensions are called Bézier surfaces. Bézier curves were widely publicized in 1962 by the French engineer Pierre Bézier, who used them to design automobile bodies. The curves were first developed in 1959 by Paul de Casteljau using de Casteljau's algorithm, a numerically stable method to evaluate Bézier curves [24].
A general Bézier curve function $B(t)$ of variable t is expressed as

$$
B(t) \;=\; \sum_{i=0}^{n} B_i \, b_i^n(t), \qquad t \in [0,1],
\tag{2.5.8}
$$

where n is a non-negative integer and B_i $(i = 0, 1, 2, \cdots, n)$ are constant coefficients; $b_i^n(t)$ is written as

$$
\begin{aligned}
b_i^n(t) &= C_i^n \, t^i \, (1 - t)^{n-i} = \binom{n}{i} t^i \, (1 - t)^{n-i}, \\
&= \frac{n!}{i!(n - i)!} \, t^i \, (1 - t)^{n-i}.
\end{aligned}
\tag{2.5.9}
$$

In a similar way, a time history of desired (d) fingertip positions (p), velocities (v), and accelerations (a) for a cubic Bézier curve function is given as

$$
\begin{aligned}
p_d^j(\hat{t}) &= B_0(1 - \hat{t})^3 + 3B_1\hat{t}(1 - \hat{t})^2 + 3B_2\hat{t}^2(1 - \hat{t}) + B_3\hat{t}^3, \\
&= B_0(1 - 3\hat{t} + 3\hat{t}^2 - \hat{t}^3) + 3B_1(\hat{t} - 2\hat{t}^2 + \hat{t}^3) + 3B_2(\hat{t}^2 - \hat{t}^3) + B_3\hat{t}^3, \\
v_d^j(\hat{t}) &= \left[B_0(-3 + 6\hat{t} - 3\hat{t}^2) + 3B_1(1 - 4\hat{t} + 3\hat{t}^2) + 3B_2(2\hat{t} - 3\hat{t}^2) + 3B_3\hat{t}^2 \right] \dot{\hat{t}}, \\
a_d^j(\hat{t}) &= \left[6B_0(1 - \hat{t}) + 6B_1(-2 + 3\hat{t}) + 6B_2(1 - 3\hat{t}) + 6B_3 \right] \dot{\hat{t}}^2.
\end{aligned}
\tag{2.5.10}
$$

Here, $B_0 - B_3$ are undetermined constants and the superscript j indicates the index of each finger, for example, $j = t, i, m, r$, and l represent thumb, index finger, middle finger, ring finger, and little finger, respectively. The transformed time $\hat{t}$ and its first derivative $\dot{\hat{t}}$ with respect to time are defined as

$$
\hat{t} = \frac{t - t_0}{t_f - t_0} \in [0, 1],
\tag{2.5.11}
$$

$$
\dot{\hat{t}} = \frac{d\hat{t}}{dt} = \frac{1}{t_f - t_0},
\tag{2.5.12}
$$

where t is the real time; t_0 and t_f are the real initial time and real terminal time, respectively. The following four constraint conditions at real time $t = t_0$ and t_f need to be satisfied.

$$
\begin{aligned}
p_d^j(\hat{t} = 0) &= p_0^j, \\
v_d^j(\hat{t} = 0) &= v_0^j, \\
p_d^j(\hat{t} = 1) &= p_f^j, \\
v_d^j(\hat{t} = 1) &= v_f^j,
\end{aligned}
\tag{2.5.13}
$$

where p_0^j and p_f^j are the initial and terminal fingertip positions of the finger j; v_0^j and v_f^j are the initial and terminal fingertip velocities of the finger j. Based on the four constraint conditions at $t = t_0$ and t_f, (2.5.10), they are rewritten as

$$
\begin{aligned}
p_d^j(\hat{t} = 0) &= B_0 = p_0^j, \\
v_d^j(\hat{t} = 0) &= \left(\frac{-3}{t_f - t_0}\right) B_0 + \left(\frac{3}{t_f - t_0}\right) B_1 = v_0^j, \\
p_d^j(\hat{t} = 1) &= B_3 = p_f^j, \\
v_d^j(\hat{t} = 1) &= \left(\frac{-3}{t_f - t_0}\right) B_2 + \left(\frac{3}{t_f - t_0}\right) B_3 = v_f^j, \quad (2.5.14)
\end{aligned}
$$

In the matrix form, the relations (2.5.14) are rewritten as

$$
\mathbf{T}^B \, \mathbf{B} \;=\; \mathbf{P}, \quad (2.5.15)
$$

where the matrices $\mathbf{T}^B$, $\mathbf{B}$, and $\mathbf{P}$ are

$$
\mathbf{T}^B \;=\; \begin{bmatrix} 1 & 0 & 0 & 0 \\ \frac{-3}{t_f - t_0} & \frac{3}{t_f - t_0} & 0 & 0 \\ 0 & 0 & 0 & 1 \\ 0 & 0 & \frac{-3}{t_f - t_0} & \frac{3}{t_f - t_0} \end{bmatrix},
$$

$$
\mathbf{B} \;=\; \begin{bmatrix} B_0 & B_1 & B_2 & B_3 \end{bmatrix}',
$$

$$
\mathbf{P} \;=\; \begin{bmatrix} p_0^j & v_0^j & p_f^j & v_f^j \end{bmatrix}'. \quad (2.5.16)
$$

Therefore, the four unknown constants, B_0–B_3, are computed by

$$
\mathbf{B} \;=\; \mathbf{T}^{B-1} \, \mathbf{P}. \quad (2.5.17)
$$

2.5.3   Simulation Results of Trajectory Paths

Figures 2.20, 2.21, and 2.22 show the trajectory position, linear velocity, and linear acceleration of fingertips of the cubic polynomial and cubic Bézier curve functions, respectively. The initial position and terminal position are 0.030 and 0.065 (m), respectively; the tracking time is 20 (sec). There is no obvious difference between the cubic polynomial and cubic Bézier curve functions based on these results. Therefore, the polynomial function will be utilized in all studied cases for control strategies. To interpolate between more than two points, Bézier curve functions are used because the control points are easily and intuitively moved around in order to adjust the curve as the user may like.

Figure 2.23 shows the forward kinematics (upper row) in terms of the thumb link angles and the tip positions and the inverse kinematics (lower row) of the thumb in terms of the tip positions and the link angles. Figure 2.24 shows forward kinematics (upper row) in terms of the index finger link angles and the tip positions and inverse kinematics (lower row) in terms of the tip positions and the link angles of the index

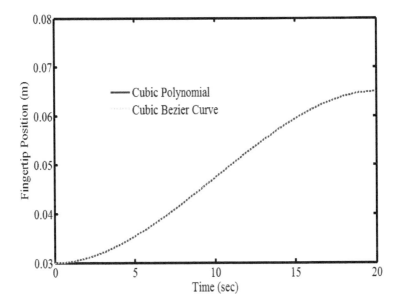

Figure 2.20   Trajectory Position of Fingertip

finger by using ANFIS (see Section 4.3) and similar results were obtained using GA (see Section 4.5). During our simulations [12], we found that GA method gives a better solution (error $\approx 10^{-7}$), but it takes more execution time whereas the ANFIS approach gave a good solution (error $\approx 10^{-4}$) with less time compared to the GA method.

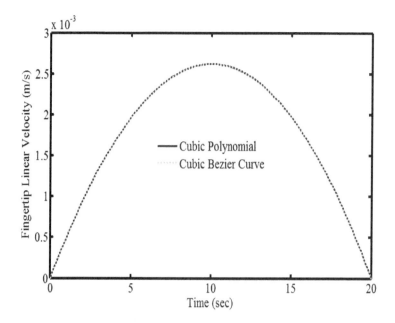

Figure 2.21   Trajectory Linear Velocity of Fingertip

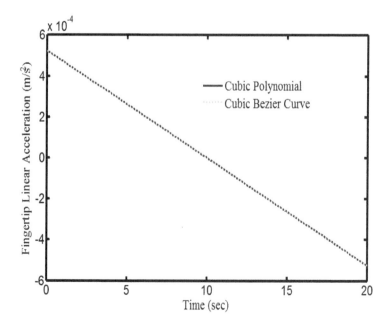

Figure 2.22   Trajectory Linear Acceleration of Fingertip

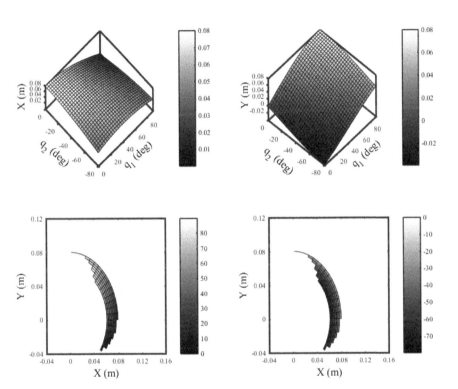

Figure 2.23   Forward Kinematics and Inverse Kinematics of Thumb

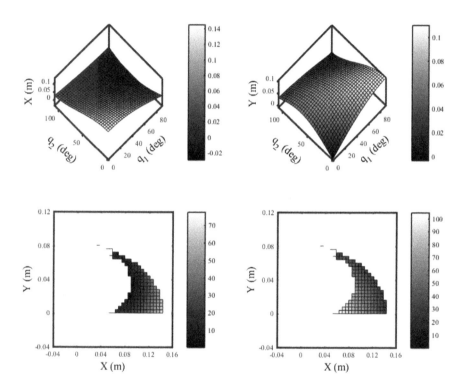

Figure 2.24 Forward Kinematics and Inverse Kinematics of Index Finger Using ANFIS

Bibliography

[1] R. R. Seeley, T. D. Stephens, and P. Tate. *Anatomy and Physiology, Eighth Edition.* The McGraw-Hill, New York, USA, 2007.

[2] C.-H. Chen, D. S. Naidu, and M. P. Schoen. Adaptive control for a five-fingered prosthetic hand with unknown mass and inertia. *World Scientific and Engineering Academy and Society (WSEAS) Journal on Systems*, 10(5):148–161, May 2011.

[3] C.-H. Chen and D. S. Naidu. Hybrid control strategies for a five-finger robotic hand. *Biomedical Signal Processing and Control*, 8(4):382–390, July 2013.

[4] C.-H. Chen. *Hybrid Control Strategies for Smart Prosthetic Hand.* PhD Dissertation, Idaho State University, Pocatello, Idaho, USA, May 2009.

[5] C.-H. Chen and D. S. Naidu. "Optimal control strategy for two-fingered smart prosthetic hand," in *Proceedings of the International Association of Science and Technology for Development (IASTED) International Conference on Robotics and Applications (RA 2010)*, pp. 190–196, Cambridge, Massachusetts, USA, November 1–3, 2010.

[6] R. J. Schilling. *Fundamentals of Robotics: Analysis and Control.* Prentice Hall, Englewood Cliffs, New Jersey, USA, 1990.

[7] R. Kelly, V. Santibanez, and A. Loria. *Control of Robot Manipulators in Joint Space.* Springer, New York, USA, 2005.

[8] M. W. Spong, S. Hutchinson, and M. Vidyasagar. *Robot Dynamics and Control.* John Wiley & Sons, New York, USA, 2006.

[9] R. N. Jazar. *Theory of Applied Robotics. Kinematics, Dynamics, and Control.* Springer, New York, USA, 2007.

[10] B. Siciliano, L. Sciavicco, L. Villani, and G. Oriolo. *Robotics: Modelling, Planning and Control.* Springer-Verlag, London, UK, 2009.

[11] C.-H. Chen, K. W. Bosworth, M. P. Schoen, S. E. Bearden, D. S. Naidu, and A. Perez-Gracia. "A study of particle swarm optimization on leukocyte adhesion molecules and control strategies for smart prosthetic hand," in *2008 IEEE Swarm Intelligence Symposium (IEEE SIS08)*, St. Louis, Missouri, USA, September 21–23, 2008.

[12] C.-H. Chen, D. S. Naidu, A. Perez-Gracia, and M. P. Schoen. "Fusion of hard and soft control techniques for prosthetic hand," in *Proceedings of the International Association of Science and Technology for Development (IASTED) International Conference on Intelligent Systems and Control (ISC 2008)*, pp. 120–125, Orlando, Florida, USA, November 16–18, 2008.

[13] C.-H. Chen, D. S. Naidu, A. Perez-Gracia, and M. P. Schoen. "A hybrid adaptive control strategy for a smart prosthetic hand," in *The 31st Annual International Conference of the IEEE Engineering Medicine and Biology Society (EMBS)*, pp. 5056–5059, Minneapolis, Minnesota, USA, September 2–6, 2009.

[14] D. G. Kamper, E. G. Cruz, and M. P. Siegel. Stereotypical fingertip trajectories during grasp. *Journal of Neurophysiology*, 90:3702–3710, 2003.

[15] X. Luo, T. Kline, H. C. Fisher, K. A. Stubblefield, R. V. Kenyon, and D. G. Kamper. "Integration of augmented reality and assistive devices for post-stroke hand opening rehabilitation," in *The International Conference of IEEE Engineering in Medicine and Biology Society (EMBS)*, Shanghai, P. R. China, 2005.

[16] L. Zollo, S. Roccella, E. Guglielmelli, M. C. Carrozza, and P. Dario. Biomechatronic design and control of an anthropomorphic artificial hand for prosthetic and robotic applications. *IEEE/ASME Transactions on Mechatronics*, 12(4):418–429, August 2007.

[17] X. Wen, D. Sheng, and J. Huang. *A Hybrid Particle Swarm Optimization for Manipulator Inverse Kinematics Control*, volume 5226 of *Lecture Notes in Computer Science*. Springer-Verlag, Berlin, Germany, 2008.

[18] C.-H. Chen, D. S. Naidu, A. Perez-Gracia, and M. P. Schoen. "A hybrid control strategy for five-fingered smart prosthetic hand," in *Joint 48th IEEE Conference on Decision and Control (CDC) and 28th Chinese Control Conference (CCC)*, pp. 5102–5107, Shanghai, P. R. China, December 16–18, 2009.

[19] C.-H. Chen, D. S. Naidu, A. Perez-Gracia, and M. P. Schoen. "A hybrid optimal control strategy for a smart prosthetic hand," in *Proceedings of the ASME 2009 Dynamic Systems and Control Conference (DSCC)*, Hollywood, California, USA, October 12–14, 2009 (No. DSCC2009–2507).

[20] J.-S. R. Jang, C.-T. Sun, and E. Mizutani. *Neuro-Fuzzy and Soft Computing: A Computational Approach to Learning and Machine Intelligence*. Prentice Hall PTR, Upper Saddle River, New Jersey, USA, 1997.

[21] P. K. Lavangie and C. C. Norkin. *Joint Structure and Function: A Comprehensive Analysis, Third Edition*. F. A. Davis Company, Philadelphia, Pennsylvania, USA, 2001.

[22] S. Arimoto. *Control Theory of Multi-fingered Hands: A Modeling and Analytical–Mechanics Approach for Dexterity and Intelligence*. Springer-Verlag, London, UK, 2008.

[23] F. L. Lewis, D. M. Dawson, and C. T. Abdallah. *Robot Manipulators Control: Second Edition, Revised and Expanded*. Marcel Dekker, Inc., New York, USA, 2004.

[24] R. H. Bartels, J. C. Beatty, and B. A. Barsky. *Bezier Curves: Chapter 10 in An Introduction to Splines for Use in Computer Graphics and Geometric Modelling*. Morgan Kaufmann, San Francisco, California, USA, 1998.

CHAPTER 3

DYNAMIC MODELS

Electric motors are used to actuate the joints of small size manipulators, like robotic hand. A transmission is the mechanical gear cables that can change the angular velocity between the electric motors and the actuated joints. Section 3.1 describes the quantitative effects by deriving the mathematical model of the actuator between direct current (DC) motor and mechanical gears. The dynamic equations of hand motion are then derived via Lagrangian approach for a serial n-link revolute-joint manipulator in Section 3.2.

3.1 Actuators

3.1.1 Electric DC Motor

For small load applications such as for a two-link thumb and three-link fingers, it is normal to use an electric DC motor as an actuator which is shown in Figure 3.1 [1]. It is easy to write the necessary relations for the DC motor circuit as

$$L\frac{di}{dt} + Ri(t) + K_b\dot{q}(t) = u(t), \qquad (3.1.1)$$

Fusion of Hard and Soft Control Strategies for the Robotic Hand, By C.-H. Chen and D. S. Naidu
© 2017 by the Institute of Electrical and Electronic Engineers, Inc. Published 2017 by John Wiley & Sons, Inc.

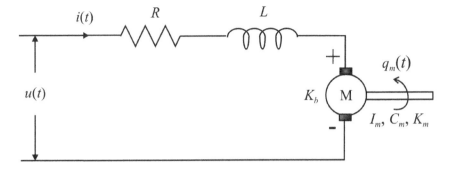

Figure 3.1 Electric DC Motor

and for the torque/force developed by the motor as

$$I_m\ddot{q}(t) + C_m\dot{q}(t) + K_m q(t) = K_b i(t), \qquad (3.1.2)$$

where, R, L, and K_b are DC motor armature resistance, inductance, back electro motive force (e.m.f.) constants, respectively, and I_m, C_m, and K_m are motor moment of inertia, damping coefficient, and spring constant, respectively. For two-link system, we need to have two actuators and, in general, we need n actuators for n-link system, in which case, we rewrite (3.1.1) and (3.1.2) in matrix form

$$\mathbf{L}\frac{d\mathbf{i}}{dt} + \mathbf{R}\mathbf{i}(t) + \mathbf{K}_b\dot{\mathbf{q}}(t) = \mathbf{u}(t), \qquad (3.1.3)$$

$$\mathbf{I}_m\ddot{\mathbf{q}}_m(t) + \mathbf{C}_m\dot{\mathbf{q}}_m(t) + \mathbf{K}_m\mathbf{q}(t) = \mathbf{K}_b\mathbf{i}(t), \qquad (3.1.4)$$

where, $\mathbf{q}(t) = \mathbf{q}_m(t)$ is the motor displacement coordinate vector, $\mathbf{I}_m \in \Re^{n \times n}$ is a diagonal and positive definite, moment-of-inertia matrix of the motors including the motor gears, $\mathbf{u}(t) \in \Re^n$ and $\mathbf{i}(t) \in \Re^n$ are vectors representing armature voltages or control inputs and currents, respectively; $\mathbf{L}, \mathbf{R}$, and $\mathbf{K}_b$ are, respectively, diagonal matrices representing armature inductances, armature resistances, and motor back e.m.f. constants.

3.1.2 Mechanical Gear Transmission

Figure 3.2 shows the transmission effects between an electric DC motor and a mechanical gear. I_m and I represent the inertia moments regarding the rotation axis for the motor and the loaded joint, respectively; F_m and F are viscous friction coefficients of the motor and the loaded joint; τ_m and τ denote the driving torque of the motor and the coupled torque of the loaded joint; q_m and q are, respectively, the angular positions of the motor and loaded joint axes. If the coupling between the motor gear and loaded joint gear is the absence of slipping, there is the relation

$$r_m q_m = r q. \qquad (3.1.5)$$

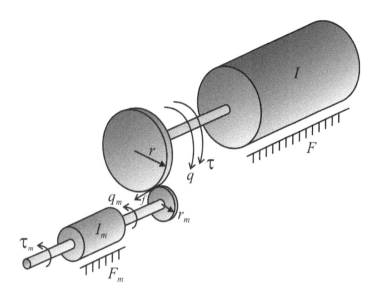

Figure 3.2　Schematic Representation of a Mechanical Gear Transmission

Here, r_m and r are the gear radii of the motor and the loaded joint, respectively. The gear reduction ratio k_r is defined as

$$k_r = \frac{r}{r_m} = \frac{q_m}{q} = \frac{\dot{q}_m}{\dot{q}} = \frac{\ddot{q}_m}{\ddot{q}}, \tag{3.1.6}$$

where $\dot{q}_m$ and $\dot{q}$ are the angular velocities of the motor and loaded joint axes, respectively; $\ddot{q}_m$ and $\ddot{q}$ represent the angular accelerations of the motor and loaded joint axes, respectively. As shown in Figure 3.2, the contact exchanged force f between the motor gear and loaded joint gear would generate a driving torque fr_m for the motor axis and a reaction torque fr for the loaded joint axis, respectively. Therefore, the moment balances for the motor axis and loaded joint axis are expressed as

$$\tau_m = I_m\ddot{q}_m + F_m\dot{q}_m + fr_m, \tag{3.1.7}$$
$$fr = I\ddot{q} + F\dot{q} + \tau. \tag{3.1.8}$$

Substituting (3.1.6) into (3.1.8) gives the relation as

$$\begin{aligned} f &= \frac{I\ddot{q} + F\dot{q} + \tau}{r}, \\ &= \left(\frac{I}{r}\right)\left(\frac{\ddot{q}_m}{k_r}\right) + \left(\frac{F}{r}\right)\left(\frac{\dot{q}_m}{k_r}\right) + \frac{\tau}{r}. \end{aligned} \tag{3.1.9}$$

Substituting (3.1.6) and (3.1.9) into (3.1.7) gives the relation as

$$
\begin{aligned}
\tau_m &= I_m \ddot{q}_m + F_m \dot{q}_m + \left[\left(\frac{I}{r} \right) \left(\frac{\ddot{q}_m}{k_r} \right) + \left(\frac{F}{r} \right) \left(\frac{\dot{q}_m}{k_r} \right) + \frac{\tau}{r} \right] r_m, \\
&= I_m \ddot{q}_m + F_m \dot{q}_m + \left(\frac{I}{k_r{}^2} \right) \ddot{q}_m + \left(\frac{F}{k_r{}^2} \right) \dot{q}_m + \frac{\tau}{k_r}, \\
&= \left(I_m + \frac{I}{k_r{}^2} \right) \ddot{q}_m + \left(F_m + \frac{I}{k_r{}^2} \right) \dot{q}_m + \frac{\tau}{k_r}.
\end{aligned}
\tag{3.1.10}
$$

(3.1.10) expresses the relation between the driving torque τ_m of the motor axis and the reaction torque τ of the loaded joint axis.

3.2 Dynamics

It is necessary to have a mathematical model that describes the dynamic behavior of robotic hand for the purpose of designing the control system. Hence, in this section the dynamic equations of hand motion are derived via Lagrangian approach using kinetic energy and potential energy.

Lagrange's equations of hand motion [2–4] are given by

$$
\begin{aligned}
\frac{d}{dt} \left(\frac{\partial \mathcal{L}}{\partial \dot{\mathbf{q}}} \right) - \frac{\partial \mathcal{L}}{\partial \mathbf{q}} &= \mathbf{Q}, \\
&= \tau - \mathbf{F_v} \dot{\mathbf{q}} - \mathbf{F_s} \mathrm{sign}(\dot{\mathbf{q}}) - \mathbf{J}' \mathbf{F_{ext}},
\end{aligned}
\tag{3.2.1}
$$

where $\mathcal{L}$ is the Lagrangian; $\dot{\mathbf{q}}$ and $\mathbf{q}$ represent the angular velocity and angle vectors of joints, respectively; $\mathbf{Q}$ is the generalized force vector; τ is the given torque vector at joints; $\mathbf{F_v}$ denotes the diagonal positive definite matrix of viscous friction coefficients. As a simplified model of static friction torques, one may consider the Coulomb friction torques $\mathbf{F_s} \mathrm{sign}(\dot{\mathbf{q}})$, where $\mathbf{F_s}$ is a diagonal positive definite matrix and $\mathrm{sign}(\dot{\mathbf{q}})$ is a vector whose components are given by the sign functions of the single joint velocities. $\mathbf{J}$ is the Jacobian and $\mathbf{F_{ext}}$ denotes the vector of external forces exerted by the end-effector on the environment. The Lagrangian $\mathcal{L}$ is expressed as

$$
\mathcal{L} = T - V,
\tag{3.2.2}
$$

where T and V denote kinetic energy and potential energy, respectively [5].

3.3 Two Link Thumb

The Lagrangian $\mathcal{L}^t$ of thumb is described as

$$
\mathcal{L}^t = T^t - V^t.
\tag{3.3.1}
$$

T^t and V^t are written as

$$
\begin{aligned}
T^t &= \sum_{k=1}^{n=2} T_k^t = \sum_{k=1}^{n=2} \left(T_k^{t,lin} + T_k^{t,rot} \right), \\
&= \sum_{k=1}^{n=2} \left(\frac{1}{2} m_k^t \mathbf{v_{ck}^t}^T \mathbf{v_{ck}^t} + \frac{1}{2} \boldsymbol{\omega}_k^{t\,T} \mathbf{I_k^t} \boldsymbol{\omega}_k^t \right), \\
&= \sum_{k=1}^{n=2} \left(\frac{1}{2} m_k^t \frac{d}{dt} \mathbf{p_{ck}^t}^T \frac{d}{dt} \mathbf{p_{ck}^t} + \frac{1}{2} \boldsymbol{\omega}_k^{t\,T} \mathbf{I_k^t} \boldsymbol{\omega}_k^t \right). \quad (3.3.2) \\
V^t &= \sum_{k=1}^{n=2} V_k^t. \quad\quad (3.3.3)
\end{aligned}
$$

Here, n is the number of links of thumb; $T_k^{t,lin}$ and $T_k^{t,rot}$ represent linear and rotational parts of kinetic energy, respectively; m_k^t denotes the mass of link k; $\mathbf{v_{ck}^t}$ is the center of mass velocity vector of link k; $\mathbf{p_{ck}^t}$ is the center of mass position vector of link k; $\boldsymbol{\omega}_k^t$ is the angular velocity vector of link k; $\mathbf{I_k^t}$ represents the moment matrix of inertia of link k; V_k^t is potential energy of link k. Here $\mathbf{p_{ck}^t}$, $\mathbf{v_{ck}^t}$, $\boldsymbol{\omega}_k^t$, $\mathbf{I_k^t}$, and V_k^t are denoted by

$$
\mathbf{p_{c1}^t} = \begin{bmatrix} l_1^t \cos(q_1^t) \\ l_1^t \sin(q_1^t) \\ 0 \end{bmatrix}, \quad
\mathbf{p_{c2}^t} = \begin{bmatrix} L_1^t \cos(q_1^t) + l_2^t \cos(q_1^t + q_2^t) \\ L_1^t \sin(q_1^t) + l_2^t \sin(q_1^t + q_2^t) \\ 0 \end{bmatrix}. \quad (3.3.4)
$$

$$
\mathbf{v_{c1}^t} = \frac{d}{dt} \mathbf{p_{c1}^t} = \frac{d}{dt} \begin{bmatrix} l_1^t \cos(q_1^t) \\ l_1^t \sin(q_1^t) \\ 0 \end{bmatrix},
$$

$$
\mathbf{v_{c2}^t} = \frac{d}{dt} \mathbf{p_{c2}^t} = \frac{d}{dt} \begin{bmatrix} L_1^t \cos(q_1^t) + l_2^t \cos(q_1^t + q_2^t) \\ L_1^t \sin(q_1^t) + l_2^t \sin(q_1^t + q_2^t) \\ 0 \end{bmatrix}. \quad (3.3.5)
$$

$$
\boldsymbol{\omega}_1^t = \frac{d}{dt} \begin{bmatrix} 0 \\ 0 \\ q_1^t \end{bmatrix} = \begin{bmatrix} 0 \\ 0 \\ \dot{q}_1^t \end{bmatrix}, \quad
\boldsymbol{\omega}_2^t = \frac{d}{dt} \begin{bmatrix} 0 \\ 0 \\ q_1^t + q_2^t \end{bmatrix} = \begin{bmatrix} 0 \\ 0 \\ \dot{q}_1^t + \dot{q}_2^t \end{bmatrix}.
$$

$$
(3.3.6)
$$

$$\mathbf{I}_1^t = \begin{bmatrix} I_{xx1}^t & -I_{xy1}^t & -I_{xz1}^t \\ -I_{yx1}^t & I_{yy1}^t & -I_{yz1}^t \\ -I_{zx1}^t & -I_{zy1}^t & I_{zz1}^t \end{bmatrix},$$

$$\mathbf{I}_2^t = \begin{bmatrix} I_{xx2}^t & -I_{xy2}^t & -I_{xz2}^t \\ -I_{yx2}^t & I_{yy2}^t & -I_{yz2}^t \\ -I_{zx2}^t & -I_{zy2}^t & I_{zz2}^t \end{bmatrix}. \tag{3.3.7}$$

$$V_1^t = m_1^t g l_1^t \sin(q_1^t), \quad V_2^t = m_2^t g L_1^t \sin(q_1^t) + m_2^t g l_2^t \sin(q_1^t + q_2^t). \tag{3.3.8}$$

Here l_k^t is the distance between the end of previous link and the center of mass of link k; L_k^t is the length of link k; q_k^t is the angle at joint k and a function of time; g is the acceleration due to gravity; the diagonal elements $I_{mnk}^t (k = 1, 2), m = n$ are called *polar moments of inertia*

$$I_{xxk}^t = I_{xk}^t = \int_{V_k} (y^2 + z^2) \, dm,$$

$$I_{yyk}^t = I_{yk}^t = \int_{V_k} (z^2 + x^2) \, dm,$$

$$I_{zzk}^t = I_{zk}^t = \int_{V_k} (x^2 + y^2) \, dm. \tag{3.3.9}$$

The off-diagonal elements $I_{mnk}^t (k = 1, 2), m \neq n$ are called *products of inertia*

$$I_{xyk}^t = I_{yxk}^t = \int_{V_k} (xy) \, dm,$$

$$I_{yzk}^t = I_{zyk}^t = \int_{V_k} (yz) \, dm,$$

$$I_{zxk}^t = I_{xzk}^t = \int_{V_k} (zx) \, dm. \tag{3.3.10}$$

Here, V_k represents the body domain of link k.

Therefore, from Lagrangian approach, dynamic equations of thumb are obtained as below [6].

$$\mathbf{M}(\mathbf{q})\ddot{\mathbf{q}} + \mathbf{C}(\mathbf{q}, \dot{\mathbf{q}}) + \mathbf{G}(\mathbf{q}) = \boldsymbol{\tau} - \mathbf{F}_v\dot{\mathbf{q}} - \mathbf{F}_s\mathbf{sign}(\dot{\mathbf{q}}) - \mathbf{J}'\mathbf{F}_{\text{ext}} \tag{3.3.11}$$

or

$$\begin{bmatrix} M_{11}^t & M_{12}^t \\ M_{21}^t & M_{22}^t \end{bmatrix} \begin{bmatrix} \ddot{q}_1^t \\ \ddot{q}_2^t \end{bmatrix} + \begin{bmatrix} C_1^t \\ C_2^t \end{bmatrix} + \begin{bmatrix} G_1^t \\ G_2^t \end{bmatrix} = \begin{bmatrix} \tau_1^t \\ \tau_2^t \end{bmatrix} - \begin{bmatrix} F_{v1}^t & 0 \\ 0 & F_{v2}^t \end{bmatrix} \begin{bmatrix} \dot{q}_1^t \\ \dot{q}_2^t \end{bmatrix}$$
$$- \begin{bmatrix} F_{s1}^t & 0 \\ 0 & F_{s2}^t \end{bmatrix} \begin{bmatrix} sign(\dot{q}_1^t) \\ sign(\dot{q}_2^t) \end{bmatrix} - \begin{bmatrix} J_{11}^t & J_{21}^t \\ J_{12}^t & J_{22}^t \end{bmatrix} \begin{bmatrix} F_x^t \\ F_y^t \end{bmatrix}. \tag{3.3.12}$$

Here,

$$
\begin{aligned}
M_{11}^t &= 2m_2^t L_1^t l_2^t \cos(q_2^t) + m_1^t l_1^{t^2} + m_2^t L_1^{t^2} + m_2^t l_2^{t^2} + I_{zz1}^t + I_{zz2}^t, \\
M_{12}^t &= m_2^t L_1^t l_2^t \cos(q_2^t) + m_2^t l_2^{t^2} + I_{zz2}^t, \\
M_{21}^t &= M_{12}^t, \\
M_{22}^t &= m_2^t l_2^{t^2} + I_{zz2}^t,
\end{aligned}
\tag{3.3.13}
$$

$$
\begin{aligned}
C_1^t &= -2m_2^t L_1^t l_2^t \sin(q_2^t)\dot{q}_1^t \dot{q}_2^t - m_2^t L_1^t l_2^t \sin(q_2^t)\dot{q}_2^t \dot{q}_2^t, \\
C_2^t &= m_2^t L_1^t l_2^t \sin(q_2^t)\dot{q}_1^t \dot{q}_1^t - m_2^t L_1^t l_2^t \sin(q_2^t)\dot{q}_1^t \dot{q}_2^t, \\
G_1^t &= g(m_1^t l_1^t \cos(q_1^t) + m_2^t L_1^t \cos(q_1^t) + m_2^t l_2^t \cos(q_1^t + q_2^t)), \\
G_2^t &= gm_2^t l_2^t \cos(q_1^t + q_2^t),
\end{aligned}
\tag{3.3.14}
$$
$$
\tag{3.3.15}
$$

τ_1^t and τ_2^t are the given torques at the joints 1 and 2, respectively; $\mathbf{M}(\mathbf{q})$ is the inertia matrix; $\mathbf{C}(\mathbf{q}, \dot{\mathbf{q}})$ is the Coriolis/centripetal vector and $\mathbf{G}(\mathbf{q})$ is the gravity vector; $\mathbf{F_v}$ denotes the diagonal positive definite matrix of viscous friction coefficients; $\mathbf{F_s}$ is a diagonal positive definite matrix and $\mathbf{sign}(\dot{\mathbf{q}})$ is a vector whose components are given by the sign functions of the single joint velocities; $\mathbf{J}$ is the Jacobian and $\mathbf{F_{ext}}$ denotes the vector of external forces in each direction. The sign function $sign(x)$ is given by

$$
sign(x) = \begin{cases} 1 & \text{if } x > 0 \\ -1 & \text{if } x < 0. \end{cases}
$$

(3.3.11) is also be written as

$$
\mathbf{M}(\mathbf{q})\ddot{\mathbf{q}} + \mathbf{N}(\mathbf{q}, \dot{\mathbf{q}}) = \tau,
\tag{3.3.16}
$$

where $\mathbf{N}(\mathbf{q}, \dot{\mathbf{q}}) = \mathbf{C}(\mathbf{q}, \dot{\mathbf{q}}) + \mathbf{G}(\mathbf{q}) + \mathbf{F_v}\dot{\mathbf{q}} + \mathbf{F_s}\mathbf{sign}(\dot{\mathbf{q}}) + \mathbf{J}'\mathbf{F_{ext}}$ represents non-linear terms in $\mathbf{q}$ and $\dot{\mathbf{q}}$.

3.4 Three Link Index Finger

Similarly, dynamic equations of index finger [6] are obtained (by software MAPLE) in the same form (3.3.11) as

$$
\begin{bmatrix} M_{11}^i & M_{12}^i & M_{13}^i \\ M_{21}^i & M_{22}^i & M_{23}^i \\ M_{31}^i & M_{32}^i & M_{33}^i \end{bmatrix} \begin{bmatrix} \ddot{q}_1^i \\ \ddot{q}_2^i \\ \ddot{q}_3^i \end{bmatrix} + \begin{bmatrix} C_1^i \\ C_2^i \\ C_3^i \end{bmatrix} + \begin{bmatrix} G_1^i \\ G_2^i \\ G_3^i \end{bmatrix} = \begin{bmatrix} \tau_1^i \\ \tau_2^i \\ \tau_3^i \end{bmatrix} - \begin{bmatrix} F_{v1}^i & 0 & 0 \\ 0 & F_{v2}^i & 0 \\ 0 & 0 & F_{v3}^i \end{bmatrix}
$$
$$
\begin{bmatrix} \dot{q}_1^i \\ \dot{q}_2^i \\ \dot{q}_3^i \end{bmatrix} - \begin{bmatrix} F_{s1}^i & 0 & 0 \\ 0 & F_{s2}^i & 0 \\ 0 & 0 & F_{s3}^i \end{bmatrix} \begin{bmatrix} sign(\dot{q}_1^i) \\ sign(\dot{q}_2^i) \\ sign(\dot{q}_3^i) \end{bmatrix} - \begin{bmatrix} J_{11}^i & J_{21}^i \\ J_{12}^i & J_{22}^i \\ 0.7J_{12}^i & 0.7J_{22}^i \end{bmatrix} \begin{bmatrix} F_x^i \\ F_y^i \end{bmatrix}.
$$
$$
\tag{3.4.1}
$$

Here,

$$
\begin{aligned}
M_{11}^i &= 2m_2^i L_1^i l_2^i \sin(q_1^i)\sin(q_1^i+q_2^i) + 2m_2^i L_1^i l_2^i \cos(q_1^i)\cos(q_1^i+q_2^i) \\
&\quad +2m_3^i L_1^i L_2^i \sin(q_1^i)\sin(q_1^i+q_2^i) \\
&\quad +2m_3^i L_1^i L_2^i \cos(q_1^i)\cos(q_1^i+q_2^i) \\
&\quad +2m_3^i L_1^i l_3^i \sin(q_1^i)\sin(q_1^i+q_2^i+q_3^i) \\
&\quad +2m_3^i L_1^i l_3^i \cos(q_1^i)\cos(q_1^i+q_2^i+q_3^i) \\
&\quad +2m_3^i L_2^i l_3^i \sin(q_1^i+q_2^i)\sin(q_1^i+q_2^i+q_3^i) \\
&\quad +2m_3^i L_2^i l_3^i \cos(q_1^i+q_2^i)\cos(q_1^i+q_2^i+q_3^i) \\
&\quad +m_1^i {l_1^i}^2 + m_2^i {L_1^i}^2 + m_2^i {l_2^i}^2 + m_3^i {L_1^i}^2 + m_3^i {L_2^i}^2 + m_3^i {l_3^i}^2 \\
&\quad +I_{zz1}^i + I_{zz2}^i + I_{zz3}^i, \\
M_{12}^i &= m_2^i L_1^i l_2^i \sin(q_1^i)\sin(q_1^i+q_2^i) + m_2^i L_1^i l_2^i \cos(q_1^i)\cos(q_1^i+q_2^i) \\
&\quad +2m_3^i L_2^i l_3^i \sin(q_1^i+q_2^i)\sin(q_1^i+q_2^i+q_3^i) \\
&\quad +2m_3^i L_2^i l_3^i \cos(q_1^i+q_2^i)\cos(q_1^i+q_2^i+q_3^i) \\
&\quad +m_3^i L_1^i L_2^i \sin(q_1^i)\sin(q_1^i+q_2^i) + m_3^i L_1^i L_2^i \cos(q_1^i)\cos(q_1^i+q_2^i) \\
&\quad +m_3^i L_1^i l_3^i \sin(q_1^i)\sin(q_1^i+q_2^i+q_3^i) \\
&\quad +m_3^i L_1^i l_3^i \cos(q_1^i)\cos(q_1^i+q_2^i+q_3^i) \\
&\quad +m_2^i {l_2^i}^2 + m_3^i {L_2^i}^2 + m_3^i {l_3^i}^2 + I_{zz2}^i + I_{zz3}^i, \\
M_{13}^i &= m_3^i L_1^i l_3^i \sin(q_1^i)\sin(q_1^i+q_2^i+q_3^i) \\
&\quad +m_3^i L_1^i l_3^i \cos(q_1^i)\cos(q_1^i+q_2^i+q_3^i) \\
&\quad +m_3^i L_2^i l_3^i \sin(q_1^i+q_2^i)\sin(q_1^i+q_2^i+q_3^i) \\
&\quad +m_3^i L_2^i l_3^i \cos(q_1^i+q_2^i)\cos(q_1^i+q_2^i+q_3^i) \\
&\quad +m_3^i {l_3^i}^2 + I_{zz3}^i,
\end{aligned} \tag{3.4.2}
$$

$$
\begin{aligned}
M_{21}^i &= M_{12}^i, \\
M_{22}^i &= 2m_3^i L_2^i l_3^i \sin(q_1^i+q_2^i)\sin(q_1^i+q_2^i+q_3^i) \\
&\quad +2m_3^i L_2^i l_3^i \cos(q_1^i+q_2^i)\cos(q_1^i+q_2^i+q_3^i) \\
&\quad +m_2^i {l_2^i}^2 + m_3^i {L_2^i}^2 + m_3^i {l_3^i}^2 + I_{zz2}^i + I_{zz3}^i, \\
M_{23}^i &= m_3^i L_2^i l_3^i \sin(q_1^i+q_2^i)\sin(q_1^i+q_2^i+q_3^i) \\
&\quad +m_3^i L_2^i l_3^i \cos(q_1^i+q_2^i)\cos(q_1^i+q_2^i+q_3^i) \\
&\quad +m_3^i {l_3^i}^2 + I_{zz3}^i,
\end{aligned} \tag{3.4.3}
$$

$$
\begin{aligned}
M_{31}^i &= M_{13}^i, \quad M_{32}^i = M_{23}^i, \\
M_{33}^i &= m_3^i {l_3^i}^2 + I_{zz3}^i.
\end{aligned} \tag{3.4.4}
$$

$$
\begin{aligned}
G_1^i &= g(m_1^i l_1^i \cos(q_1^i) + m_2^i L_1^i \cos(q_1^i) + m_3^i L_1^i \cos(q_1^i) \\
&\quad +m_1^i l_2^i \cos(q_1^i+q_2^i) + m_3^i L_2^i \cos(q_1^i+q_2^i) \\
&\quad +m_3^i l_3^i \cos(q_1^i+q_2^i+q_3^i)), \\
G_2^i &= g(m_2^i l_2^i \cos(q_1^i+q_2^i) + m_3^i L_2^i \cos(q_1^i+q_2^i) \\
&\quad +m_3^i l_3^i \cos(q_1^i+q_2^i+q_3^i)), \\
G_3^i &= g(m_3^i l_3^i \cos(q_1^i+q_2^i+q_3^i)).
\end{aligned} \tag{3.4.5}
$$

$$
\begin{aligned}
C_1^i =\ & (2m_2^i L_1^i l_2^i \sin(q_1^i)\cos(q_1^i+q_2^i) - 2m_2^i L_1^i l_2^i \cos(q_1^i)\sin(q_1^i+q_2^i) \\
& +2m_3^i L_1^i L_2^i \sin(q_1^i)\cos(q_1^i+q_2^i) - 2m_3^i L_1^i L_2^i \cos(q_1^i)\sin(q_1^i+q_2^i) \\
& +2m_3^i L_1^i l_3^i \sin(q_1^i)\cos(q_1^i+q_2^i+q_3^i) - 2m_3^i L_1^i l_3^i \cos(q_1^i)\sin(q_1^i+q_2^i+q_3^i)) \times \\
& \left(\frac{\partial q_1^i}{\partial t}\right)\left(\frac{\partial q_2^i}{\partial t}\right) \\
& +(2m_3^i L_1^i l_3^i \sin(q_1^i)\cos(q_1^i+q_2^i+q_3^i) - 2m_3^i L_1^i l_3^i \cos(q_1^i)\sin(q_1^i+q_2^i+q_3^i) \\
& +2m_3^i L_2^i l_3^i \sin(q_1^i+q_2^i)\cos(q_1^i+q_2^i+q_3^i) - 2m_3^i L_2^i l_3^i \cos(q_1^i+q_2^i)\sin(q_1^i+q_2^i+q_3^i)) \times \\
& \left(\frac{\partial q_1^i}{\partial t}\right)\left(\frac{\partial q_3^i}{\partial t}\right) \\
& +(2m_3^i L_1^i l_3^i \sin(q_1^i)\cos(q_1^i+q_2^i+q_3^i) - 2m_3^i L_1^i l_3^i \cos(q_1^i)\sin(q_1^i+q_2^i+q_3^i) \\
& +2m_3^i L_2^i l_3^i \sin(q_1^i+q_2^i)\cos(q_1^i+q_2^i+q_3^i) - 2m_3^i L_2^i l_3^i \cos(q_1^i+q_2^i)\sin(q_1^i+q_2^i+q_3^i)) \times \\
& \left(\frac{\partial q_2^i}{\partial t}\right)\left(\frac{\partial q_3^i}{\partial t}\right) \\
& +(m_2^i L_1^i l_2^i \sin(q_1^i)\cos(q_1^i+q_2^i) - m_2^i L_1^i l_2^i \cos(q_1^i)\sin(q_1^i+q_2^i) \\
& +m_3^i L_1^i L_2^i \sin(q_1^i)\cos(q_1^i+q_2^i) - m_3^i L_1^i L_2^i \cos(q_1^i)\sin(q_1^i+q_2^i) \\
& +m_3^i L_1^i l_3^i \sin(q_1^i)\cos(q_1^i+q_2^i+q_3^i) - m_3^i L_1^i l_3^i \cos(q_1^i)\sin(q_1^i+q_2^i+q_3^i)) \times \\
& \left(\frac{\partial q_2^i}{\partial t}\right)\left(\frac{\partial q_2^i}{\partial t}\right) \\
& +(m_3^i L_1^i l_3^i \sin(q_1^i)\cos(q_1^i+q_2^i+q_3^i) - m_3^i L_1^i l_3^i \cos(q_1^i)\sin(q_1^i+q_2^i+q_3^i) \\
& +m_3^i L_2^i l_3^i \sin(q_1^i+q_2^i)\cos(q_1^i+q_2^i+q_3^i) - m_3^i L_2^i l_3^i \cos(q_1^i+q_2^i)\sin(q_1^i+q_2^i+q_3^i)) \times \\
& \left(\frac{\partial q_3^i}{\partial t}\right)\left(\frac{\partial q_3^i}{\partial t}\right), \\[6pt]
C_2^i =\ & (m_2^i L_1^i l_2^i \sin(q_1^i)\cos(q_1^i+q_2^i) - m_2^i L_1^i l_2^i \cos(q_1^i)\sin(q_1^i+q_2^i) \\
& +m_3^i L_1^i L_2^i \sin(q_1^i)\cos(q_1^i+q_2^i) - m_3^i L_1^i L_2^i \cos(q_1^i)\sin(q_1^i+q_2^i) \\
& +m_3^i L_1^i l_3^i \sin(q_1^i)\cos(q_1^i+q_2^i+q_3^i) - m_3^i L_1^i l_3^i \cos(q_1^i)\sin(q_1^i+q_2^i+q_3^i)) \times \\
& \left(\frac{\partial q_1^i}{\partial t}\right)\left(\frac{\partial q_2^i}{\partial t}\right) \\
& +(2m_3^i L_2^i l_3^i \sin(q_1^i+q_2^i)\cos(q_1^i+q_2^i+q_3^i) - 2m_3^i L_2^i l_3^i \cos(q_1^i+q_2^i)\sin(q_1^i+q_2^i+q_3^i)) \times \\
& \left(\frac{\partial q_1^i}{\partial t}\right)\left(\frac{\partial q_3^i}{\partial t}\right) \\
& +(2m_3^i L_2^i l_3^i \sin(q_1^i+q_2^i)\cos(q_1^i+q_2^i+q_3^i) - 2m_3^i L_2^i l_3^i \cos(q_1^i+q_2^i)\sin(q_1^i+q_2^i+q_3^i)) \times \\
& \left(\frac{\partial q_2^i}{\partial t}\right)\left(\frac{\partial q_3^i}{\partial t}\right) \\
& +(-m_2^i L_1^i l_2^i \sin(q_1^i)\cos(q_1^i+q_2^i) + m_2^i L_1^i l_2^i \cos(q_1^i)\sin(q_1^i+q_2^i) \\
& -m_3^i L_1^i L_2^i \sin(q_1^i)\cos(q_1^i+q_2^i) + m_3^i L_1^i L_2^i \cos(q_1^i)\sin(q_1^i+q_2^i) \\
& -m_3^i L_1^i l_3^i \sin(q_1^i)\cos(q_1^i+q_2^i+q_3^i) + m_3^i L_1^i l_3^i \cos(q_1^i)\sin(q_1^i+q_2^i+q_3^i)) \times \\
& \left(\frac{\partial q_1^i}{\partial t}\right)\left(\frac{\partial q_1^i}{\partial t}\right) \\
& +(m_3^i L_2^i l_3^i \sin(q_1^i+q_2^i)\cos(q_1^i+q_2^i+q_3^i) - m_3^i L_2^i l_3^i \cos(q_1^i+q_2^i)\sin(q_1^i+q_2^i+q_3^i)) \times \\
& \left(\frac{\partial q_3^i}{\partial t}\right)\left(\frac{\partial q_3^i}{\partial t}\right),
\end{aligned}
$$

$$C_3^i = (2m_3^i L_2^i l_3^i \cos(q_1^i + q_2^i) \sin(q_1^i + q_2^i + q_3^i) - 2m_3^i L_2^i l_3^i \sin(q_1^i + q_2^i) \cos(q_1^i + q_2^i + q_3^i)) \times$$

$$\left(\frac{\partial q_1^i}{\partial t}\right)\left(\frac{\partial q_2^i}{\partial t}\right)$$

$$+(m_3^i L_1^i l_3^i \sin(q_1^i) \cos(q_1^i + q_2^i + q_3^i) - m_3^i L_1^i l_3^i \cos(q_1^i) \sin(q_1^i + q_2^i + q_3^i)$$

$$+(m_3^i L_2^i l_3^i \sin(q_1^i + q_2^i) \cos(q_1^i + q_2^i + q_3^i) - m_3^i L_2^i l_3^i \cos(q_1^i + q_2^i) \sin(q_1^i + q_2^i + q_3^i)) \times$$

$$\left(\frac{\partial q_1^i}{\partial t}\right)\left(\frac{\partial q_3^i}{\partial t}\right)$$

$$+(m_3^i L_2^i l_3^i \sin(q_1^i + q_2^i) \cos(q_1^i + q_2^i + q_3^i) - m_3^i L_2^i l_3^i \cos(q_1^i + q_2^i) \sin(q_1^i + q_2^i + q_3^i)) \times$$

$$\left(\frac{\partial q_2^i}{\partial t}\right)\left(\frac{\partial q_3^i}{\partial t}\right)$$

$$+(m_3^i L_1^i l_3^i \cos(q_1^i) \sin(q_1^i + q_2^i + q_3^i) - m_3^i L_1^i l_3^i \sin(q_1^i) \cos(q_1^i + q_2^i + q_3^i)$$

$$+m_3^i L_2^i l_3^i \cos(q_1^i + q_2^i) \sin(q_1^i + q_2^i + q_3^i) - m_3^i L_2^i l_3^i \sin(q_1^i + q_2^i) \cos(q_1^i + q_2^i + q_3^i)) \times$$

$$\left(\frac{\partial q_1^i}{\partial t}\right)\left(\frac{\partial q_1^i}{\partial t}\right)$$

$$+(m_3^i L_2^i l_3^i \cos(q_1^i + q_2^i) \sin(q_1^i + q_2^i + q_3^i) - m_3^i L_2^i l_3^i \sin(q_1^i + q_2^i) \cos(q_1^i + q_2^i + q_3^i)) \times$$

$$\left(\frac{\partial q_2^i}{\partial t}\right)\left(\frac{\partial q_2^i}{\partial t}\right). \tag{3.4.6}$$

Bibliography

[1] B. C. Kuo. *Automatic Control Systems, Seventh Edition*. Prentice Hall, Engle-wood Cliffs, New Jersey, USA, 1995.

[2] R. Kelly, V. Santibanez, and A. Loria. *Control of Robot Manipulators in Joint Space*. Springer, New York, USA, 2005.

[3] R. N. Jazar. *Theory of Applied Robotics. Kinematics, Dynamics, and Control*. Springer, New York, USA, 2007.

[4] B. Siciliano, L. Sciavicco, L. Villani, and G. Oriolo. *Robotics: Modelling, Planning and Control*. Springer-Verlag, London, UK, 2009.

[5] C.-H. Chen. *Hybrid Control Strategies for Smart Prosthetic Hand*. PhD Dissertation, Idaho State University, Pocatello, Idaho, USA, May 2009.

[6] C.-H. Chen, D. S. Naidu, and M. P. Schoen. Adaptive control for a five-fingered prosthetic hand with unknown mass and inertia. *World Scientific and Engineering Academy and Society (WSEAS) Journal on Systems*, 10(5):148–161, May 2011.

CHAPTER 4

SOFT COMPUTING/CONTROL STRATEGIES

The hard control strategies offer *fixed* (or *crisp*) solutions to control dynamical systems. A knowledge-based system is more proper to complete tasks where the nature of the problems and solutions is not well defined or not known [1]. This chapter introduces some soft computing (SC) or computational intelligence (CI) [2] strategies involving fuzzy logic (FL) in Section 4.1, neural network (NN) in Section 4.2, adaptive neuro-fuzzy inference system (ANFIS) in Section 4.3, tabu search (TS) in Section 4.4, genetic algorithm (GA) in Section 4.5, particle swarm optimization (PSO) in Section 4.6, developed adaptive particle swarm optimization (APSO) in Section 4.7, and condensed hybrid optimization (CHO) in Section 4.8. Then, all simulation results and discussion are presented in Section 4.9.

4.1 Fuzzy Logic

Humans are flexible and can adapt to unfamiliar situations and they can get information in an efficient manner and discard irrelevant details. The information generated does not need to be complete or precise, but it may be general, qualitative, approximate, and fuzzy for human beings, so it can reason, infer, and deduce new

Fusion of Hard and Soft Control Strategies for the Robotic Hand, By C.-H. Chen and D. S. Naidu

information and knowledge [1]. A classical (crisp) set A of real numbers larger than 30 are written as

$$A = \{x \mid x < 30\}. \tag{4.1.1}$$

Here, there is a boundary number 30. If x is smaller than this number, then x belongs to the set A; otherwise, x does not belong to the set A. Let x and A be someone's age and young, respectively. If Chen's age (x) is 29.99, which is smaller than 30 and belongs to the set A, then people may consider Chen is young (A); however, if Mario's age is 30.01, which is larger than 30 and does not belong to the set A, in real life, people do not think Mario is not young. In other words, the difference between 29.99 and 30.01 is not obvious. In contrast to a classical set, a **fuzzy set** is a set without a sharp (crisp) boundary or without binary characteristics. In other words, the identification between "belongs to the set" and "does not belong to the set" is gradual or gradient. As shown in Figures 4.1 and 4.2, such statements as "Chen

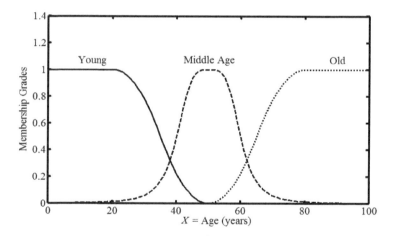

Figure 4.1   Membership Functions of "Mario is Young"

is good" and "Mario is young" are simple examples. Fuzzy sets are mathematical objects modeling this impreciseness and use the concept of degrees of **membership function** to give a mathematical definition of fuzzy sets. If X is a collection denoted by x, then a fuzzy set A in X is defined as a set of ordered pairs

$$A = \{(x, \mu_A(x)) \mid x \in X \cap \mu_A(x) \in [0, 1]\}. \tag{4.1.2}$$

Here, the membership function (MF) $\mu_A(x)$ represents the grade of *possibility* that an element x belongs to the fuzzy set A. Figures 4.1 and 4.2 utilize three membership functions: Z-shape $\mu_z(x)$, generalized bell curve $\mu_g(x)$, and S-shape $\mu_s(x)$ to present young/bad, middle, and old/good, respectively. The three membership

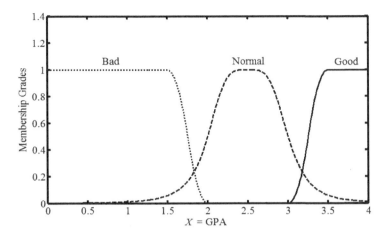

Figure 4.2   Membership Functions of "Chen is Good"

functions are expressed as

$$
\mu_z(x) \;=\; \begin{cases} 1 & \text{if } x \le a_z \\ 1 - 2\left(\frac{x-a_z}{b_z-a_z}\right)^2 & \text{if } a_z \le x \le \frac{a_z+b_z}{2} \\ 2\left(\frac{b_z-x}{b_z-a_z}\right)^2 & \text{if } \frac{a_z+b_z}{2} \le x \le b_z \\ 0 & \text{if } x \ge b_z, \end{cases}
$$

$$
\mu_g(x) \;=\; \frac{1}{1 + \left|\frac{x-c_g}{a_g}\right|^{2b_g}},
$$

$$
\mu_s(x) \;=\; \begin{cases} 0 & \text{if } x \le a_s \\ 2\left(\frac{x-a_s}{b_s-a_s}\right)^2 & \text{if } \frac{a_s+b_s}{2} \le x \le b_s \\ 1 - 2\left(\frac{b_s-x}{b_s-a_s}\right)^2 & \text{if } a_s \le x \le \frac{a_s+b_s}{2} \\ 1 & \text{if } x \ge b_s. \end{cases}
$$

Here, the parameters a_z, b_z, a_g, b_g, c_g, a_s, and b_s are selected constants.

Furthermore, humans do many things that can be classified as control by the rule "IF..., THEN...". For example, if the temperature is low, then turn on the heat; if you are tired, then get some sleep. A **fuzzy IF–THEN rule** (or named **fuzzy rule/fuzzy implication**) describes the form "IF x is A, THEN y is B."

4.2 Neural Network

A biological neuron is the foundation of all ANNs, hence forth simply called NNs. The biological neurons consists of cell body, dendrites, synapses, and axon [3]. The axon is considered as a long tube that divides into branches. The end branches of the neuron's axon terminate in synaptic knobs. Therefore, the axon of a single neuron forms synaptic connections with many other neurons. The synaptic knob of the first neuron is separated from the receptor site of another neuron by a microscopic distance. The cell body of the first neuron produces chemical materials called neurotransmitters, which are delivered down to the synaptic vesicles. The neurotransmitters are stored in the synaptic vesicles until the neuron fires and a burst of neurotransmitters is released by the vesicles. The neurotransmitters then flow across the synaptic cleft and act on the second neuron. Hence, the neurotransmitters of the first neuron may stimulate or inhibit the second neuron activity. Each neuron is receiving messages from hundreds or thousands of other neurons.

That is to say that the dendrites receive signals (electrical action potential) from other neurons through synaptic connections and the signals travel to cell body where they are received, integrated, and transmitted. The cell body of a neuron adds the incoming signals from dendrites. If the strengths of input signals reach the threshold level, then a particular neuron will fire and send a signal to its axon; if the input signals do not reach the required threshold level, then the input signals will quickly decay and will not generate any action on the axon [3].

Inspired by biological nervous systems, NN is typically composed of a set of parallel and distributed processing units, called nodes or neurons. These are usually ordered into layers, appropriately interconnected by means of unidirectional weighted signal channels, called connections or synaptic weights [1, 4–6]. All or parts of nodes are adaptive. In other words, the outputs of the nodes modifies the parameters related to these nodes. The learning rule of NN updates these parameters to minimize a prescribed error measure, which is the difference between the network's actual output and a desired output. Figure 4.3 shows a simple feedforward ANN with input layer, hidden layer, and output layer.

4.3 Adaptive Neuro Fuzzy Inference System

In 1965, L. A. Zadeh first proposed a fuzzy set, which is characterized by a membership function [7]. Unlike a crisp set, the fuzzy set defines a grade of membership value between zero and one. Using the fussy set, FL provides a mathematical model to describe the uncertain environment and approximate knowledge reasoning. Fuzzy logic controller (FLC) is developed by the combination of the fuzzy set and fuzzy logic. The FLC uses fuzzy "IF-THEN" *Rule Base* (or named fuzzy rule/fuzzy implication) to control systems. The "IF" part is known as *antecedent* and the "THEN" part is called as *consequent* [1, 8]. Among the FLC approaches, the two most common methodologies are Mamdani [9] model and Sugeno (or Takagi–Sugeno–Kang, TSK) model [10]. The two models use the same approaches to fuzzify crisp inputs

Input Layer   Hidden Layer   Output Layer

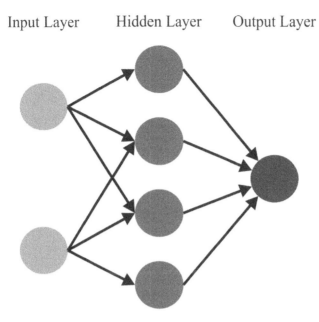

Figure 4.3   A Simple Feedforward Artificial Neural Network

and apply fuzzy operator, but the main difference between Mamdani and Sugeno is that Sugeno type uses either linear or constant (e.g. first-order or zero-order polynomial functions) for outputs without using membership functions.

A NN is a model structure with an algorithm, which can learn, train, and adjust weight parameters to fit given nonlinear data. Inspired by biological nervous systems, NN is typically composed of a set of parallel and distributed processing units (called nodes or neurons). These nodes are usually ordered into layers (including input, hidden and output layers), which are appropriately interconnected by means of unidirectional weighted signal channels (named connections or synaptic weights) [1, 4, 11]. All or partial nodes adjust the weight parameters using learning rules and algorithms, such as least squared error (LSE, also known as Widrow–Hoff learning rule) [12], gradient decent algorithm etc. In other words, the outputs of the nodes adapt the weight parameters relating to these nodes and the learning rules update these parameters to either minimize a prescribed error measure, which is the difference between the network's actual output and a desired/targeted output, or change a learning rate for the gradient of the prescribed error measure. The NN uses the adaptive LSE algorithm to fit the given data with highly nonlinear systems and the gradient decent algorithm to control convergent speed.

FL systems implement fuzzy sets to model uncertainty and approximate knowledge reasoning, but FL architecture lacks learning rules. NN systems strengthen adaptive learning rules for numerical sets to fit nonlinear data, but NN feature lacks knowledge representation. Neuro-fuzzy systems (NFS) include intelligent systems which combine the main features of both FL and NN systems to solve problems that cannot be solved with desired performance by using either FL or NN methodology alone [13]. The most common NFS is ANFIS [14]. ANFIS is a fuzzy inference system embedded in the framework of adaptive networks which provides the best optimization algorithm for finding parameters to fit the given data. Based on human reasoning in the form of fuzzy "IF-THEN" rules, ANFIS develops the mapping of input and output data pairs using a hybrid learning procedure.

The analytical solutions of inverse kinematics for two-link thumb can be deduced mathematically. However, with more complex structures (such as increasing DOF in three-dimensional space), it will become a difficult problem to solve. Since the forward kinematics for three-link fingers are formulated [15], the workspace of fingertips in Cartesian coordinates is developed by the entire range of rotating angles of all links. Therefore, the inverse kinematics problems for three-link fingers can be solved by using ANFIS [4, 14] with the Cartesian space as inputs and the joint space as outputs. The Cartesian space (inputs) and joint space (outputs) are stored as training data set and then trained by ANFIS. ANFIS includes *premise parameters* (or named *antecedent parameters*), defining membership functions, and *consequent parameters*, determining the coefficients of each output equation. The hybrid learning procedure of ANFIS uses the backpropagation gradient descent algorithm in backward pass to tune the premise parameters of membership functions and LSE (also known as Widrow–Hoff learning rule) [12] method in forward pass to adjust the consequent parameters of output functions.

J.-S. Jang presented three types of ANFIS [14] and we use type-3 ANFIS, which uses Takagi–Sugeno's fuzzy "IF-THEN" rules whose outputs are a linear combination of input variables and a constant. To simply summarize the type-3 ANFIS architecture, we assume that x and y are two input variables and $f(x, y)$ is one output variable. As shown in Figure 4.4(a), for a first-order Sugeno fuzzy model, the two Takagi–Sugeno's fuzzy "IF-THEN" rules are expressed as

Rule 1: IF (x is A_1) and (y is B_1), THEN ($f_1 = p_1 x + q_1 y + r_1$),
Rule 2: IF (x is A_2) and (y is B_2), THEN ($f_2 = p_2 x + q_2 y + r_2$).

Here, A_i and B_i ($i = 1, 2$) are the linguistic label (like *small*, *medium*, and *large*) and p_i, q_i, and r_i ($i = 1, 2$) are the linear consequent parameters. Figure 4.4(b) depicts the corresponding equivalent ANFIS architecture with five layers, named as Fuzzification (Layer 1), Product (Layer 2), Normalization (Layer 3), Defuzzification (Layer 4), and Aggregation (Layer 5) [16, 17]. The inverse kinematics problems for three-link fingers are solved by ANFIS with the Cartesian space as inputs and the joint space as outputs. The Cartesian space (inputs) and joint space (outputs) are stored as training data set and then trained by ANFIS. The hybrid learning procedure of ANFIS uses the backpropagation gradient descent algorithm in backward pass to tune the premise/antecedent parameters of membership functions (Layer 1) and the LSE method in forward pass to adjust the consequent parameters of output functions (Layer 4) [4, 14]. The structure of these layers is described below.

Layer 1: Fuzzification Layer

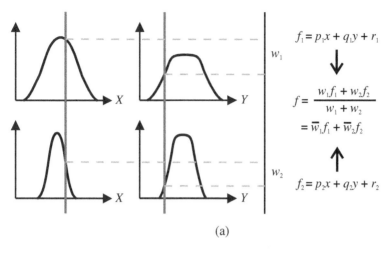

(a)

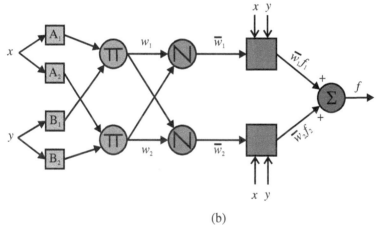

(b)

Figure 4.4 ANFIS Architecture: (a) A First Order Sugeno Fuzzy Model with Two Input Variables x and y and One Output Variable f Based on Two Fuzzy "IF THEN" Rules. (b) Corresponding Equivalent ANFIS Structure.

We define that $O_{i,j}$ is the output of the jth node in the ith layer $(i, j \in \aleph)$ and square nodes $(i = 1, 4)$ are adaptive nodes with parameters, while circle nodes $(i = 2, 3, 5)$ are fixed nodes without parameters. Nodes $j \ (= 1, 2)$ in this layer $i \ (= 1)$ are square (adaptive) nodes with node functions

$$
\begin{aligned}
O_{1,j} &= \mu_{A_j}(x), \\
O_{1,j} &= \mu_{B_j}(y),
\end{aligned}
\tag{4.3.1}
$$

where the two crisp inputs x and y to the nodes j are fuzzified through membership functions μ_{A_j} and μ_{B_j} of the linguistic labels correlated to the node functions $O_{1,j}$. The commonly used membership functions are triangular, trapezoid, Gaussian-shaped, and bell-shaped membership functions $\in [0, 1]$. For example, the bell-shaped membership function is given by

$$\mu_{A_j}(x) = \frac{1}{1 + \left[\left(\frac{x - c_j}{a_j} \right)^2 \right]^{b_j}} \tag{4.3.2}$$

where the parameter set $\{a_j, b_j, c_j\}$ includes *premise/antecedent parameters*.

Layer 2: Product (T-norm Operation) Layer
Nodes j (= 1, 2) in this layer i (= 2) are circle (fixed) nodes labeled *product* (or T-norm $\otimes$) operator $\prod$ with node functions $O_{2,j}$. The nodes in this layer multiply all incoming signals and send the product outputs to next layer (Layer 3), which represent the firing strength of fuzzy antecedent rules ("IF" part). The outputs in this layer acts as weight functions w_j and is expressed as

$$O_{2,j} = w_j = \mu_{A_j}(x) \otimes \mu_{B_j}(y). \tag{4.3.3}$$

Layer 3: Normalization Layer
Nodes j (= 1, 2) in this layer i (= 3) are circle nodes labeled **N** with node functions $O_{3,j}$. The jth node in this layer calculates the ratio of the jth rule's firing strength to the sum of all rules' firing strengths. The outputs in this layer normalize the weight functions that are transmitted from the previous product layer and the normalized weight functions (firing strengths) $\bar{w}_j$ are written as

$$O_{3,j} = \bar{w}_j = \frac{w_j}{\sum_j w_j}. \tag{4.3.4}$$

Layer 4: Defuzzification (Consequent) Layer
Nodes j (= 1, 2) in this layer i (= 4) are square nodes with node functions $O_{4,j}$. The jth node in this layer defuzzifies the fuzzy consequent rule ("THEN" part). The defuzzified outputs in this layer are multiplied by normalized firing strengths based on the formulation

$$O_{4,j} = \bar{w}_j f_j = \bar{w}_j(p_j x + q_j y + r_j). \tag{4.3.5}$$

Here, the parameter set $\{p_j, q_j, r_j\}$ consists of *consequent parameters*.

Layer 5: Aggregation (Summation) Layer
The single node in this layer i (= 5) is a circle node labeled $\sum$ with a node function $O_{5,1}$. The output in this layer calculates the total overall output as the summation of all incoming signals and is expressed as

$$O_{5,1} = \sum_j \bar{w}_j f_j = \frac{\sum_j w_j f_j}{\sum_j w_j}. \tag{4.3.6}$$

4.4 Tabu Search

Several algorithms can be used to find a global minimum in optimization problems. Among these methods, tabu search (TS), genetic algorithm (GA), and PSO are very common evolutionary algorithms. TS, developed by F. Glover, was basically a set of concepts [18, 19] used to optimize combinatorial optimization problems. These concepts were extended to solve continuous optimization problems known as continuous tabu search (CTS), which was introduced by P. Siarry and G. Berthiau [20]. Enhanced continuous tabu search (ECTS) was an algorithm that uses advanced concepts of tabu search such as diversification and intensification for optimizing functions of continuous variables. ECTS was introduced by R. Chelouh and P. Siarry [21].

4.4.1 Tabu Concepts

In this section, the definition and concepts used in TS are explained first before introducing ECTS algorithm. Figure 4.5 shows a set of all possible solutions which can

Tabu Balls

Promising Balls

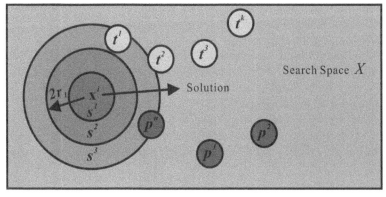

Figure 4.5 Characterization of Search Space for Enhanced Continuous Tabu Search

be visited during the search operation and is called the search space [22]. The dimension of the search space is equal to the number of variables in the cost function. The search space is denoted as X and any elements in the search space is represented by $\mathbf{x}^i$. For functions of n variables, the ambient search space is $\Re^n$ and the ith element $\mathbf{x}^i = (x_1^i, x_2^i, \ldots, x_n^i)$.

The cost function or objective function of the element $\mathbf{x}^i$ is denoted as $f(\mathbf{x}^i)$. For a function of n variables, $f(\mathbf{x}^i)$ is mapping from $\Re^n \rightarrow \Re^1$. The objective of the algorithm is to minimize

$$f(\mathbf{x}^i) : \mathbf{x}^i \in X. \tag{4.4.1}$$

For each point in the search space X, one can create a set of neighbors, which form a subspace in the search space X, the neighborhood space, denoted as $S \subset X$. Each element j in S is denoted as $\mathbf{s}^j = (s_1^j, s_2^j, \ldots, s_n^j)$, where n is the number of variables or the dimension of the search space X.

The key ingredient in TS is tabu list. In order to prevent the repetition of movements in the search space X (cycling), one uses an array, which is called tabu list (TL). TL records the most recent movement so that the recorded position will not be visited in the future. The recorded movement in TL is called "Tabu," which was inspired by a taboo, meaning a strong social prohibition/ban against words, objects, actions, or discussions that are considered undesirable or offensive by a group, culture, society, or community. The number of iterations of the Tabu is known as tabu tenure (TT), which depends on the length of the TL (N_1). The larger the number N_1, the longer the TT. N_1 plays an important role in the performance of TS algorithm. Each element k in TL is denoted as $\mathbf{t}^k = (t_1^k, t_2^k, \ldots, t_n^k)$.

4.4.2 Enhanced Continuous Tabu Search

Figure 4.6 shows the flowchart of ECTS. ECTS uses the advanced concepts of TS, namely *Diversification* and *Intensification*. ECTS is composed of four stages: initialization of parameters, diversification, selecting the most promising area, and intensification. In the following, each step is explained in details.

4.4.2.1 Initialization of Parameters   The parameters initialized in ECTS include the length of TL (N_1), the length of promising list (PL) (N_2), the radius of the neighborhood (r_1), the radius of tabu balls (r_2), the radius of the promising balls (r_3), and a random point in the search space X.

4.4.2.2 Diversification   In this stage, the algorithm looks for the most promising areas in the search space X. The step-by-step procedure for this stage is given as follows:

1. Generation of homogeneous neighbors: As shown in Figure 4.5, to any point $\mathbf{x}^i \subset X$, $\mathbf{x}^i$ generates N neighbors $\mathbf{s}^j$ such that

$$(j - 1) \, r_1 \leq ||\mathbf{x}^i - \mathbf{s}^j||_2 \leq j \, r_1. \tag{4.4.2}$$

Here $j = 1, 2, \ldots, N$; r_1 is the initial radius of the neighborhood and

$$||\mathbf{x}^i - \mathbf{s}^j||_2 = \sqrt{(x_1^i - s_1^j)^2 + \ldots + (x_n^i - s_n^j)^2}. \tag{4.4.3}$$

The above method partitions the search space into concentric spheres.

2. Comparison with the tabu list (TL): Each neighbor $\mathbf{s}^j$ ($j = 1, 2, \ldots, N$) generated in the previous step is compared with the elements in the TL $\mathbf{t}^k$ ($k = 1, 2, \ldots, N_1$) and if

$$||\mathbf{t}^k - \mathbf{s}^j||_2 \leq r_2, \qquad (4.4.4)$$

where

$$||\mathbf{t}^k - \mathbf{s}^j||_2 = \sqrt{(t_1^k - s_1^j)^2 + \ldots + (t_n^k - s_n^j)^2}, \qquad (4.4.5)$$

then the corresponding $\mathbf{s}^j$ is rejected as tabu.

3. Comparison with the promising list (PL): Let N' be the number of neighbors which are not in the TL and are denoted as $\mathbf{h}^m$ ($m = 1, 2, \ldots, N'$). Each of these elements are compared with the elements in the PL denoted as $\mathbf{p}^n$ ($n = 1, 2, \ldots, N_2$) and if

$$||\mathbf{p}^n - \mathbf{h}^m||_2 \leq r_3, \qquad (4.4.6)$$

where

$$||\mathbf{p}^n - \mathbf{h}^m||_2 = \sqrt{(p_1^n - h_1^m)^2 + \ldots + (p_n^n - h_n^m)^2}, \qquad (4.4.7)$$

then the corresponding $\mathbf{s}^j$ is rejected as Tabu.

4. Finding the best neighbor: From the neighbors which are not in the TL, one finds the best neighbor which has the minimum cost function.

5. Updating the tabu list: The best neighbor obtained is updated into the TL in a first in first out (FIFO) fashion.

6. Updating the promising list: If the best neighbor obtained is the overall best, then it is updated into the PL in a FIFO fashion.

7. Conversion determination: If no improvement occurs for a certain number of movements, then the diversification part is terminated.

4.4.2.3 *Selecting the Most Promising Area*   The most promising area from the PL is selected by two approaches, constant radius and standard deviation.

1. Constant radius: The upper bound (PA_{ub}) and lower bound (PA_{lb}) of the most promising region is the promising list (PL) $\pm$ the constant radius (r_b) divided by the dimension n and are expressed as

$$PA_{ub} = \frac{PL + r_b}{n}, \qquad (4.4.8)$$

$$PA_{lb} = \frac{PL - r_b}{n}. \qquad (4.4.9)$$

2. Standard deviation: The upper bound (PA_{ub}) and lower bound (PA_{lb}) of the most promising area care the promising list (PL) $\pm$ the standard deviation (SD) of the fixed number simulations divided by the dimension n and are expressed as

$$PA_{ub} = \frac{PL + SD}{n}, \qquad (4.4.10)$$

$$PA_{lb} = \frac{PL - SD}{n}. \qquad (4.4.11)$$

4.4.2.4 Intensification   The most promising area from the PL is selected and the above steps 1 through 7 are repeated in order to intensify the search in the most promising area.

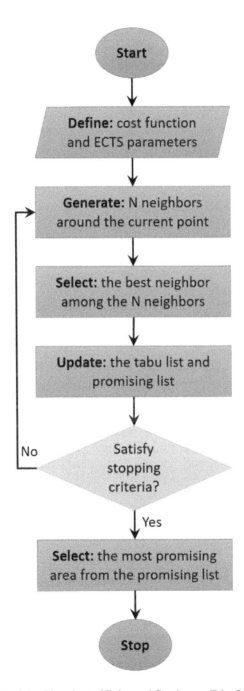

Figure 4.6   Flowchart of Enhanced Continuous Tabu Search

4.5 Genetic Algorithm

In 1859, Charles Darwin (1809–1882) published a famous book "On the origin of species by means of natural selection," which is now known as *The Origin of Species*. This is called Darwin's Theory. He suggested that in almost all organisms, there is a huge potential for the production of offspring, like eggs, but that only a small percentage survive to adulthood. To face the randomly variable environment, the living things have to change their characteristics in order to survive. Therefore, evolution is as the natural selection of inheritable variations.

Around the same time, Gregor Mendel (1822–1884) investigated the inheritance of characteristics, or traits, in his experiments with pea plants. These experiments supported Darwin's Theory. However, 30 years after Mendel's death, Walter Sutton (1877–1916) discovered that the genes of fruit flies were part of chromosomes in the nucleus. It means that if a characteristic is determined by a single gene, mutation may have a dramatic effect; if a buck of genes combines to control that characteristic, mutation in one of them may only have an unimportant effect.

The idea of GA was inspired from the chromosomes of living things which had to change their characteristics in order to survive in a randomly varying environment. Therefore, GA is a stochastic search and optimization method based on the metaphors of natural biological evolution and represented by some operators, such as **selection, crossover**, and **mutation**. GA applied the survival principles of the fittest, reproduction, and mutation to successively produce the good approximate solutions, so GA was used to solve combinatorial optimization problems. The concepts of GA were extended to solve continuous optimization problems, yielding techniques known as continuous genetic algorithms (CGA) which were developed by P. Siarry and G. Berthiau [23]. CGA is similar to GA except that the parameters are coded in terms of continuous numbers whereas in GA parameters are coded in binary format. The employed CGA uses a linear ranking model for the chromosome selection, where the probability density function is generated based on the cost values of the individual candidate chromosomes.

4.5.1 Basic GA Procedures

Figure 4.7 shows the flowchart of GA [24]. The procedure is briefly stated below.

1. Define the CGA parameters: Including initial population (Ipop), population at the end of the first generation (pop), number of chromosomes kept for mating (Keep), mutation rate (Mut), tolerance ϵ, and so on.

2. Create a homogeneous population: Generate N elements (chromosomes) and N is the Ipop.

3. Evaluate cost (fitness) function of each chromosome: Calculate the fitness value f_i of the ith member in the population.

4. Select mate based on the performance of each gene: Create a new population from the current population based on the ranking of the current fitness value,

for example, determine which parents participate in producing offspring for the next generation.

5. Reproduce the generation by crossover: Use the single or multiple crossover points to generate new chromosomes that retain the good feature and discard the bad feature.

6. Mutate: Utilize the Mut which can randomly mutate the gene to avoid falling into the local optimal area.

7. Repeat steps 3–6 until it reaches the maximum number of iterations or stopping condition defined by ϵ is satisfied.

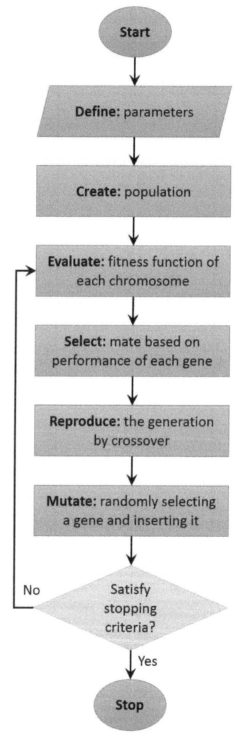

Figure 4.7   Flowchart of Genetic Algorithm

4.6 Particle Swarm Optimization

In 1995, J. Kennedy, a social psychologist, and R. Eberhart, developed a new evolutionary algorithm—PSO [25]. They implemented mathematical operators inspired from the social behaviors of bird flocks and fish schools [26]. The initialization process of PSO was similar to GA by utilizing a random population. However, GA used crossover and mutation to update the chromosomes after each generation. Unlike GA, PSO adopted the velocities, local best positions, and global best solutions of particles to renew the solutions [25–30]. Compared with other evolutionary optimization methods, PSO employed only a few simple rules in response to complex behaviors, so it was computationally inexpensive in terms of memory requirements and less time consuming than GA. Many researchers have successfully proven the benefit of PSO in different problem settings [31–45]. We first provide an overview of how basic PSO works, including procedures and formulations, and then investigate the PSO dynamics by five different techniques and uniform and random distributions [46].

In 2003, PSO researchers generally classified PSO as bridging five areas: algorithms, topology, parameters, emerging with other evolutionary computational methods, and applications [47]. Among the categories, parameters are the most important. Clerc and Kennedy investigated the explosion, stability, and convergence by constriction factors [36]. Shi and Eberhart developed an inertia weight method and compared the two ways: constriction factors and inertia weight method [34]; they also studied the selection of parameters in PSO [33]. However, utilizing few parameters and retaining the numerical stability and accuracy of the algorithm are the most important objectives. Therefore, the study of PSO dynamics is presented here.

4.6.1 Basic PSO Procedures and Formulations

PSO has the feature of being a simple and easy process as shown in Figure 4.8 [46]. The procedure is stated below.

1. Define the input parameters of PSO: Including the maximum number of iterations, swarm size, the limiting velocity (V_{max}), and upper bound (hi) and lower bound (lo) of positions for the search space, and the stopping tolerance $\epsilon > 0$.

2. Initialize particle positions: This is done randomly by either the use of a uniform distribution or a normal distribution [46].

3. Evaluate the quality of fitness function f (objective function) for the jth particle: If the new particle position can produce the better cost value, then the new positions replace the old positions, for example, update the local best and global best cost and position values. Moreover, consider the tolerance ϵ and compare the difference $|f^*(t) - f^*(t-1)|$ between the current (t) and previous $(t-1)$ best cost values. If $|f^*(t) - f^*(t-1)| < \epsilon$, then stop the loop; otherwise, go to the next step.

4. As shown in Figure 4.9, updating the velocity $V_i^j(t)$ and position $x_i^j(t)$ vectors is accomplished by the equation of motion:

$$
\begin{aligned}
V_i^j(t) &= \alpha(t)V_i^j(t-1) + \beta(t)[x_i^{j,lbest}(t-1) - x_i^j(t-1)] \\
&\quad + \gamma(t)[x_i^{j,gbest}(t-1) - x_i^j(t-1)], \qquad (4.6.1) \\
x_i^j(t) &= x_i^j(t-1) + V_i^j(t)\Delta t, \qquad\qquad\qquad\qquad (4.6.2)
\end{aligned}
$$

where $V_i^j(t)$ and $V_i^j(t-1)$ are the i component velocity of the j particle in the time t and $t-1$, respectively; $\alpha(t)$, $\beta(t)$, and $\gamma(t)$ are random values drawn from a uniform distribution or normal distribution, which will be explained in next Section 4.6.2; $lbest$ and $gbest$ represent local best and global best positions in each generation; Δt is the increment time for each iteration, for example, $\Delta t = 1$ in each iteration. (4.6.1) and (4.6.2) are expressed in the matrix form:

$$
\begin{bmatrix} V_i^j(t) \\ x_i^j(t) \end{bmatrix} = \begin{bmatrix} \alpha(t) & -\beta(t) - \gamma(t) \\ \alpha(t) & 1 - \beta(t) - \gamma(t) \end{bmatrix} \begin{bmatrix} V_i^j(t-1) \\ x_i^j(t-1) \end{bmatrix}
$$
$$
+ \begin{bmatrix} \beta(t) & \gamma(t) \\ \beta(t) & \gamma(t) \end{bmatrix} \begin{bmatrix} x_i^{j,lbest}(t-1) \\ x_i^{j,gbest}(t-1) \end{bmatrix}. \qquad (4.6.3)
$$

Note, if the velocity values of the particles are too large, they will make the particles leave the search space too often [26–28]. Hence, to avoid the explosion of particles, a limiting velocity constrain V_{max} is necessary [26], for example, if the velocity values $V_i^j(t) > V_{max}$, then $V_i^j(t) = V_{max}$; if $V_i^j(t) < -V_{max}$, $V_i^j(t) = -V_{max}$. Similarly, the boundary conditions are considered, for example, if the updated positions $x_i^j(t) > hi$, then $x_i^j(t) = hi$; if the updated positions $x_i^j(t) < lo$, then $x_i^j(t) = lo$, where the suffix i is the component of particle dimension, the superscript j represents the index of particles, and t means time.

5. Repeat steps 2–4 until it reaches the maximum number of iterations or stopping condition defined by ϵ is satisfied.

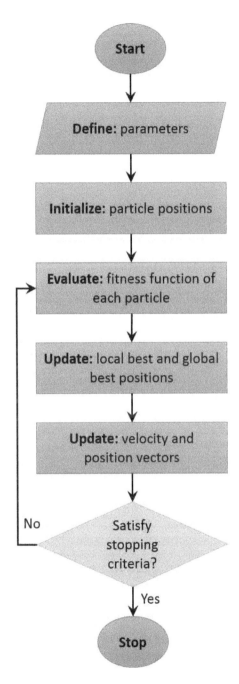

Figure 4.8   Flowchart of Particle Swarm Optimization

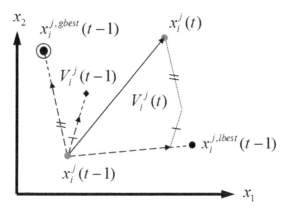

Figure 4.9   Illustration of Previous and Updated Position of Particles

4.6.2 Five Different PSO Techniques

According to (4.6.1)–(4.6.3), the updating velocities and positions are based on local best positions and global best position in each iteration. There are five techniques to determine the next velocities and positions by the local best and global best.

1. Swarm's local best: In this technique, each updating local best cost is expressed by the swarm's local best position of each iteration [26], as shown in Figure 4.10.

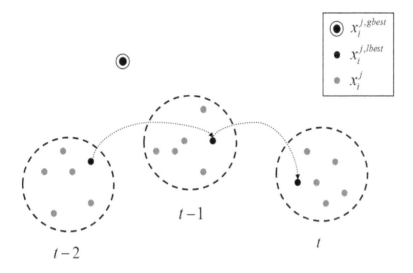

Figure 4.10   Illustration of Each Movement of Swarm's Local Best

2. Particle's local best: Figures 4.11 and 4.12 compare the two different local best techniques in terms of fitness cost [28].

3. Swarm's local best without global best cost: This approach is simply using technique A, but ignoring the global best value.

4. Swarm's local best with spline weighting function: As shown in Figure 4.9, $V_i^j(t)$ is determined by the addition of $V_i^j(t-1)$, $x_i^{j,lbest}(t-1) - x_i^j(t-1)$, and $x_i^{j,gbest}(t-1) - x_i^j(t-1)$ vectors times a random value. However, if the j particle is close to swarm's local best or global best position, the influence of $x_i^{j,lbest}(t-1) - x_i^j(t-1)$ and $x_i^{j,gbest}(t-1) - x_i^j(t-1)$ should be smaller. Conversely, if the particle is far away to the swarm's local best or global best, the influence of $x_i^{j,lbest}(t-1) - x_i^j(t-1)$ and $x_i^{j,gbest}(t-1) - x_i^j(t-1)$ should be increased. After considering the continuity and numerical stability of the velocity, a spline function is selected as the modified weighting functions $W_i^{j,lbest}$

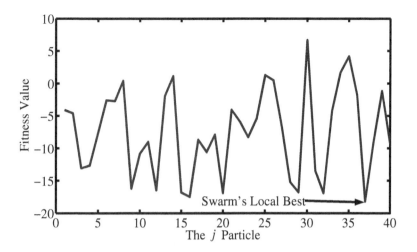

Figure 4.11   Illustration of Swarm's Local Best

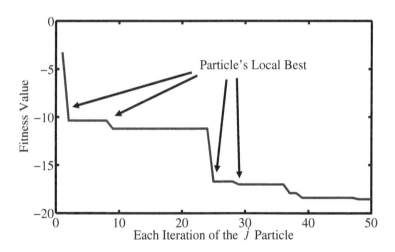

Figure 4.12   Illustration of Particle's Local Best

and $W_i^{j,gbest}$ [48], given as follows:

$$W_i^{j,lbest} = \begin{cases} 6\left(\frac{d_i^{j,lbest}}{r_i}\right)^2 - 8\left(\frac{d_i^{j,lbest}}{r_i}\right)^3 + 3\left(\frac{d_i^{j,lbest}}{r_i}\right)^4 & \text{if } 0 \leq d_i^{j,lbest} \leq r_i, \\ 1 & \text{if } r_i \leq d_i^{j,lbest}, \end{cases}$$

and

$$W_i^{j,gbest} = \begin{cases} 6\left(\frac{d_i^{j,gbest}}{r_i}\right)^2 - 8\left(\frac{d_i^{j,gbest}}{r_i}\right)^3 + 3\left(\frac{d_i^{j,gbest}}{r_i}\right)^4 & \text{if } 0 \leq d_i^{j,gbest} \leq r_i, \\ 1 & \text{if } r_i \leq d_i^{j,gbest}, \end{cases}$$

where $d_i^{j,lbest} = |x_i^j - x_i^{j,lbest}|$ and $d_i^{j,gbest} = |x_i^j - x_i^{j,gbest}|$, respectively; $d_i^{j,lbest}$ is the distance between the j particle positions x_i^j and the local best positions $x_i^{j,lbest}$; $d_i^{j,gbest}$ is the distance between the j particle positions x_i^j and the global best positions $x_i^{j,gbest}$; r_i is the radius of the support domains, as shown in Figure 4.13. Therefore, (4.6.1) and (4.6.3) are rewritten as

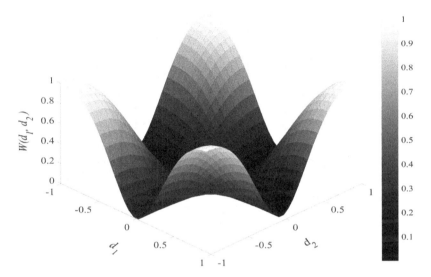

Figure 4.13   Spline Weighting Function

$$\begin{aligned} V_i^j(t) &= \alpha(t)V_i^j(t-1) + W_i^{j,lbest}\beta(t)(x_i^{j,lbest}(t-1) - x_i^j(t-1)) \\ &+ W_i^{j,gbest}\gamma(t)(x_i^{j,gbest}(t-1) - x_i^j(t-1)), \end{aligned} \tag{4.6.4}$$

and

$$\begin{bmatrix} V_i^j(t) \\ x_i^j(t) \end{bmatrix} = \begin{bmatrix} \alpha(t) & -\beta(t)W_i^{j,lbest} - \gamma(t)W_i^{j,gbest} \\ \alpha(t) & 1 - \beta(t)W_i^{j,lbest} - \gamma(t)W_i^{j,gbest} \end{bmatrix} \begin{bmatrix} V_i^j(t-1) \\ x_i^j(t-1) \end{bmatrix}$$

$$+ \begin{bmatrix} \beta(t)W_i^{j,lbest} & \gamma(t)W_i^{j,gbest} \\ \beta(t)W_i^{j,lbest} & \gamma(t)W_i^{j,gbest} \end{bmatrix} \begin{bmatrix} x_i^{j,lbest}(t-1) \\ x_i^{j,gbest}(t-1) \end{bmatrix}. \tag{4.6.5}$$

5. Swarm's local best with spline weighting function and the gradient of fitness function f: The technique of conjugate gradient has the feature of fast convergence in GA [49]. We introduce a modified descent method mimicking conjugate gradient. The updating positions, (4.6.2), of the modified gradient with

spline weighting technique are rewritten as

$$x_i^j(t) = x_i^j(t-1) + V_i^j(t)\Delta t - \delta(t)W_i^{j,gbest}\frac{\nabla_i f}{||f||_2}, \qquad (4.6.6)$$

where $\delta(t)$ is a random variable, which will be explained in details in the next Section 4.6.3; $||f||_2$ is the norm of the fitness function f and $\nabla_i f$ is the gradient of the fitness function f in the i direction. (4.6.3) is also rewritten as

$$\begin{bmatrix} V_i^j(t) \\ x_i^j(t) \end{bmatrix} = \begin{bmatrix} \alpha(t) & -\beta(t)W_i^{j,lbest} - \gamma(t)W_i^{j,gbest} \\ \alpha(t) & 1 - \beta(t)W_i^{j,lbest} - \gamma(t)W_i^{j,gbest} \end{bmatrix} \begin{bmatrix} V_i^j(t-1) \\ x_i^j(t-1) \end{bmatrix}$$
$$+ \begin{bmatrix} \beta(t)W_i^{j,lbest} & \gamma(t)W_i^{j,gbest} \\ \beta(t)W_i^{j,lbest} & \gamma(t)W_i^{j,gbest} \end{bmatrix} \begin{bmatrix} x_i^{j,lbest}(t-1) \\ x_i^{j,gbest}(t-1) \end{bmatrix}$$
$$- \begin{bmatrix} 0 \\ \delta(t)W_i^{j,gbest}\frac{\nabla_i f}{||f||_2} \end{bmatrix}. \qquad (4.6.7)$$

4.6.3   Uniform Distribution and Normal Distribution

In (4.6.1)–(4.6.7), $\alpha(t)$, $\beta(t)$, $\gamma(t)$, and $\delta(t)$ are random variables based on the uniform distribution or normal distribution. The variable's values based on the uniform distribution, donated as Uni, are $\in [0,1]$. However, the variable's value based on the normal distribution is shown as $N(\mu,\sigma)$, where μ is the mean value and σ is standard deviation; the general formula for the probability density function (pdf) with normal distribution is given as follows:

$$f(x) = \frac{exp\left(\frac{-1}{2}\left(\frac{x-\mu}{\sigma}\right)^2\right)}{\sigma\sqrt{2\pi}}. \qquad (4.6.8)$$

Figure 4.14 shows each normal distribution. Here, N(0,1), N(1,2), N(1,1), N(1,1/2), and N(1,1/3) are indicated as Nor1, Nor2a, Nor2b, Nor2c, and Nor2d, respectively.

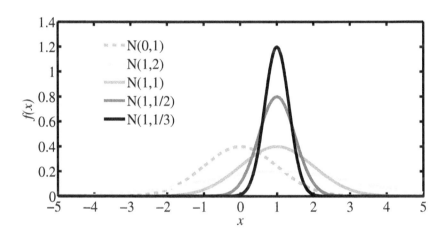

Figure 4.14 Illustration of Each Normal Distribution

4.7 Adaptive Particle Swarm Optimization

In Step 4 of the basic PSO procedure in Section 4.6.1, the maximum velocity (V_{max}) is utilized to avoid the explosion of particles. However, for multi-dimensional problems, the V_{max} should be adaptive in order to get the better performance. Hence, we proposed an adaptive PSO (APSO) method to change this V_{max} [50].

4.7.1 APSO Procedures and Formulations

The proposed APSO first develops the linear model by selecting $n + 1$ points from the swarm and then calculate the relative error standard deviation by selecting the remaining points from the swarm. Eventually, fuzzy logic rules based on the computed relative error standard deviation and the tilt of the developed linear plane are utilized to adapt the limited velocity V_{max}. The procedure is detailed below.

1. Construct a linear model: Figure 4.15 expresses the illustration of swarm's po-

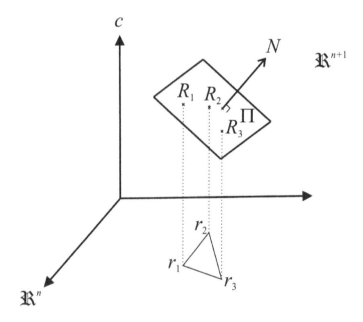

Figure 4.15 Illustration of Swarm's Positions and Cost in n Dimension

sitions and cost in an n-dimensional space. Vectors in $\Re^n$ are $\mathbf{r}$ and vectors in $\Re^{n+1}$ are $\mathbf{R}$, so the position vectors $\mathbf{R}_j$ and $\mathbf{r}_j$ of the jth particle are described as

$$\mathbf{r}_j = \left(x_1^j(t), x_2^j(t), \cdots, x_i^j(t) \right), \tag{4.7.1}$$

$$\mathbf{R}_j = (\mathbf{r}_j, c_j). \tag{4.7.2}$$

Here, $j = 1, 2, \cdots, Ss$; Ss is the swarm size; $i = n$, which is the size of the multi-dimensional space; c_j represents the cost value of the jth particle.

By randomly selecting $n + 1$ particles from the swarm size Ss, a plane Π is created as

$$\mathbf{N} \cdot \mathbf{R} \ = \ b, \tag{4.7.3}$$

which fits these points. Here, $\mathbf{N} = (N_1, N_2, \cdots, N_{n+1})$ represents the normal vector of the created plane Π. $N_1, N_2, \cdots, N_{n+1}$ are non-zero components of $\mathbf{N}$. Note that $\mathbf{N}$ and b are not uniquely defined (any multiple of both describes the same plane), but we must ensure that the linear model's "plane" never will be vertical. Therefore, we require that $N_{n+1} = 1$. (4.7.3) is rewritten as

$$\mathbf{R} \cdot \mathbf{N} - b \ = \ 0, \tag{4.7.4}$$

for $\mathbf{R} = \mathbf{R}_1, \mathbf{R}_2, \cdots, \mathbf{R}_{n+1}$ with $N_{n+1} = 1$. We rearrange (4.7.4) in the matrix form as

$$\mathbf{Au} \ = \ \mathbf{w}, \tag{4.7.5}$$

where

$$\mathbf{A} = \begin{bmatrix} \mathbf{r}_1 & c_1 & -1 \\ \mathbf{r}_2 & c_2 & -1 \\ \vdots & \vdots & \vdots \\ \mathbf{r}_{n+1} & c_{n+1} & -1 \\ 0 & 1 & 0 \end{bmatrix}, \tag{4.7.6}$$

$$\mathbf{u} = \begin{bmatrix} N_1 \\ N_2 \\ \vdots \\ N_{n+1} \\ b \end{bmatrix} = \begin{bmatrix} N_1 \\ N_2 \\ \vdots \\ 1 \\ b \end{bmatrix}, \quad \mathbf{w} = \begin{bmatrix} 0 \\ 0 \\ \vdots \\ 0 \\ 1 \end{bmatrix}. \tag{4.7.7}$$

Hence, $\mathbf{u}$ is solved by $\mathbf{u} = \mathbf{A}^{-1}\mathbf{w}$. Now we have the linear equations describing the cost surface as determined by the choice of $\mathbf{r}_1, \cdots, \mathbf{r}_{n+1}$, and $c_1, \cdots, c_{n+1}$. The cost surface is approximated by

$$\mathbf{N} \cdot \mathbf{R} \ = \ b,$$

$$(N_1, \cdots, N_n, N_{n+1}) \cdot \mathbf{R} \ = \ b,$$

or

$$(N_1, \cdots, N_n, 1) \cdot (\mathbf{r}, c) \;=\; b. \tag{4.7.8}$$

Here $\mathbf{r} = (x_1, x_2, \cdots, x_n) \in \Re^n$. Then we solve for c.

$$c = -(N_1, \cdots, N_n) \cdot \mathbf{r} + b = L(\mathbf{r}). \tag{4.7.9}$$

$L(\mathbf{r})$ is the calculated linear model for the cost (fitness) function $f(\mathbf{r})$, for example, $L(\mathbf{r}) \approx f(\mathbf{r})$.

2. Calculate the relative error standard deviation and the tilt: We use $L(\mathbf{r})$ to compute the errors (deviations in the cost surface to the linear fit), and subsequently the standard deviation. Select randomly k points (k is selected by $k = Ss - (n + 1)$) from the swarm size Ss and compare the vectors as given below.

$$
\begin{aligned}
(\tilde{\mathbf{r}}_1, \tilde{c}_1) \;&=: \; \tilde{\mathbf{R}}_1, \\
(\tilde{\mathbf{r}}_2, \tilde{c}_2) \;&=: \; \tilde{\mathbf{R}}_2, \\
&\;\;\vdots \\
(\tilde{\mathbf{r}}_k, \tilde{c}_k) \;&=: \; \tilde{\mathbf{R}}_k.
\end{aligned}
\tag{4.7.10}
$$

Thus, the errors e_i are defined as

$$e_i \;:=\; \tilde{c}_i - L(\tilde{\mathbf{r}}_i), \tag{4.7.11}$$

for $i = 1, 2, \cdots, k$. $\tilde{c}_i$ is the cost value of the particle i.
Substitute (4.7.9) into (4.7.11) and obtain the error e_i from

$$
\begin{aligned}
e_i \;&=\; \tilde{c}_i - L(\tilde{\mathbf{r}}_i), \\
&=\; (N_1, \cdots, N_n) \cdot \mathbf{r}_i + \tilde{c}_i - b.
\end{aligned}
\tag{4.7.12}
$$

Accordingly, the error array $\mathbf{e}$ is computed as

$$
\mathbf{e} =
\begin{bmatrix}
e_1 \\
e_2 \\
\vdots \\
e_{k-1} \\
e_k
\end{bmatrix}
=
\begin{bmatrix}
\tilde{\mathbf{r}}_1 & \tilde{c}_1 & 1 \\
\tilde{\mathbf{r}}_2 & \tilde{c}_2 & 1 \\
\vdots & \vdots & \vdots \\
\tilde{\mathbf{r}}_{k-1} & \tilde{c}_{k-1} & 1 \\
\tilde{\mathbf{r}}_k & \tilde{c}_k & 1
\end{bmatrix}
\begin{bmatrix}
N_1 \\
\vdots \\
N_n \\
1 \\
-b
\end{bmatrix}
=
\begin{bmatrix}
\tilde{\mathbf{R}} & 1
\end{bmatrix}
\tilde{\mathbf{u}}, \tag{4.7.13}
$$

where

$$
\tilde{\mathbf{R}} =
\begin{bmatrix}
\tilde{\mathbf{r}}_1 & \tilde{c}_1 \\
\tilde{\mathbf{r}}_2 & \tilde{c}_2 \\
\vdots & \vdots \\
\tilde{\mathbf{r}}_{k-1} & \tilde{c}_{k-1} \\
\tilde{\mathbf{r}}_k & \tilde{c}_k
\end{bmatrix},
\quad
\tilde{\mathbf{u}} =
\begin{bmatrix}
N_1 \\
\vdots \\
N_n \\
1 \\
-b
\end{bmatrix}.
\tag{4.7.14}
$$

Consequently, the error standard deviation S_k of the sample (the selected k points) is

$$
\begin{aligned}
S_k &= \sqrt{\frac{1}{k-1}\sum_{i=1}^{k}\left(\tilde{c}_i - L(\tilde{\mathbf{r}}_i)\right)^2}, \\
&= \sqrt{\frac{1}{k-1}\sum_{i=1}^{k}e_i{}^2}.
\end{aligned}
\tag{4.7.15}
$$

The cost standard deviation of the swarm is denoted as S_n, so the relative error standard deviation S_{rel} is given as

$$
S_{rel} = \frac{S_k}{S_n}.
\tag{4.7.16}
$$

Next, the tilt of the linear plane Π is computed. The normalized normal vectors $\mathbf{N}_s$ of the plane Π are calculated as

$$
\mathbf{N}_s = \frac{|\mathbf{N}'|}{||\mathbf{N}'||_2} = \frac{|\mathbf{N}'|}{\sqrt{\mathbf{N}'\mathbf{N}}} \in [0,1].
\tag{4.7.17}
$$

Here, $\mathbf{N} = [N_1\ N_2\ \cdots\ N_{n+1}]$. "Small" tilt means $\mathbf{N}_s$ near 1, while "large" tilt means $\mathbf{N}_s$ near zero.

3. Fuzzy logic rules and membership functions: As shown in Figure 4.16, four fuzzy logic rules are utilized as stated below.

(1) IF $\mathbf{N}_s$ is "small" tilt AND S_{rel} is "large," THEN "decrease" velocity scale S_v.

(2) IF $\mathbf{N}_s$ is "small" tilt AND S_{rel} is "small," THEN compare the cost and the cost of the last "flat" area. IF the current cost is smaller than the cost corresponding to the previous flat area, THEN "increase" velocity scale S_v. However, IF the current cost is larger than the cost corresponding to the previous flat area, THEN "decrease" velocity scale S_v.

(3) IF $\mathbf{N}_s$ is "big" tilt AND S_{rel} is "small," THEN "increase" velocity scale S_v.

(4) IF $\mathbf{N}_s$ is "big" tilt AND S_{rel} is "large," THEN "decrease" velocity scale S_v.

Two membership functions are used in this work: Z-shape and S-shape. The Z-shaped and S-shaped membership functions $f_z(x)$ and $f_s(x)$ of x are respectively described as

$$
f_z(x) = \begin{cases}
1 & \text{if } x \leq a_z, \\
1 - 2\left(\frac{x - a_z}{b_z - a_z}\right)^2 & \text{if } a_z \leq x \leq \frac{a_z + b_z}{2}, \\
2\left(\frac{b_z - x}{b_z - a_z}\right)^2 & \text{if } \frac{a_z + b_z}{2} \leq x \leq b_z, \\
0 & \text{if } x \geq b_z,
\end{cases}
$$

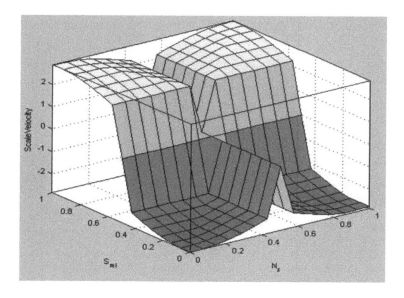

Figure 4.16   Fuzzy Logic Rule Surface of the Proposed APSO

and

$$
f_s(x) = \begin{cases} 0 & \text{if } x \le a_s, \\ 2\left(\frac{x-a_s}{b_s-a_s}\right)^2 & \text{if } \frac{a_s+b_s}{2} \le x \le b_s, \\ 1 - 2\left(\frac{b_s-x}{b_s-a_s}\right)^2 & \text{if } a_s \le x \le \frac{a_s+b_s}{2}, \\ 1 & \text{if } x \ge b_s. \end{cases}
$$

Here, the parameters a_z, b_z, a_s, and b_s are selected and given in Table 4.1. $\mathbf{N}_s$ and S_{rel} are input variables while S_v is an output variable. Therefore, the adaptive maximum velocity V_{max} is computed from

$$
V_{max} = \exp(S_v). \tag{4.7.18}
$$

4.7.2   Changed/Unchanged Velocity Direction

In Step 4 of the basic PSO procedure (Section 4.6.1), the V_{max} constrains the updating velocity. However, the updating velocity directions are changed as shown in Figure 4.17(a). We propose an unchanged updating velocity direction approach. As shown in Figure 4.17(b), the updating velocity direction is unchanged by the multiple of the updating velocity unit vector and V_{max}.

$$
\mathbf{V}(t) = \frac{\mathbf{V}(t-1)}{||\mathbf{V}(t-1)||_2} V_{max}. \tag{4.7.19}
$$

Table 4.1   Selection of Membership Function Parameters

Variables	Changed	Z/S-Shapes	a_z/a_s	b_z/b_s
$\mathbf{N}_s$	Big	Z	0	0.5
$\mathbf{N}_s$	Small	S	0.5	1
S_{rel}	Small	Z	0	0.5
S_{rel}	Large	S	0.5	1
S_v	Increase	Z	-4	0
S_v	Decrease	S	0	4

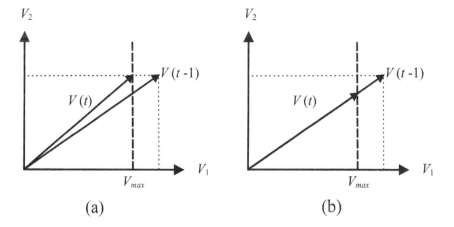

Figure 4.17   Updating Velocity Approaches by (a) Changed and (b) Unchanged Updating Velocity Directions

4.8 Condensed Hybrid Optimization

TS and PSO have some inherent advantages and disadvantages. TS is capable of searching wide regions in the search space, but is not guaranteed to fall into the global optimum solutions. However, the experimental results showed that TS did find good solutions near global optimum [51]. PSO was inherently slow but gave better convergence than TS to achieve the optimal solutions. The performance of PSO could be enhanced by providing some additional information, such as search space. The disadvantages of TS and PSO could be mitigated by combining them; such attempts had been done for combinatorial optimization problems [52, 53]. Our previous works [54, 55] presented the hybrid algorithm which combined ECTS and CGA to solve two- and three-dimensional continuous optimization problems and parameter estimation in the presence of colored noise. Nonetheless, the implementation of the GA operators was complicated, required large amount of memory, and consumed a lot of CPU time.

The proposed CHO algorithm consists of the diversification portion, using an ECTS algorithm, and an intensification portion, which is represented by a PSO instead of an ECTS [22]. In other words, one intensifies the search in the most promising area of the search space using a PSO. The flow chart of the proposed CHO algorithm is depicted in Figure 4.18. The diversification and the intensification are from ECTS and PSO, respectively.

The diversification of CHO contains the six steps below.

1. Define: Cost function and ECTS parameters

2. Generate: N neighbors around the current point $\mathbf{x}^i \subset X$ to the tabu list nor the promising list

3. Select: The best neighbor among the N neighbors and make it as the new current point

4. Update: The tabu list and promising list

5. Stop: The inner loop will stop when it reaches one of the termination criteria, which is defined as

   (a) Terminate after a fixed number of iterations (the maximum iteration I_{max})

   (b) After a certain number of iterations without any improvement in the cost function, for example, all decreases are smaller than the tolerance ϵ

   (c) When the objective function reaches a pre-specified value

6. Submit: The selected most promising area from the promising list

The intensification of CHO includes another six steps below.

1. Define: Cost function and PSO parameters

2. Evaluate: Cost function of each particle

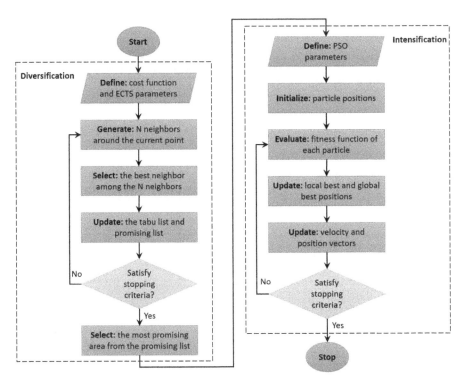

Figure 4.18 Diversification and Intensification Loops of Proposed CHO Algorithm

3. Update: The local best and global best positions

4. Renew: The velocity and position vectors of each particle

5. Stop: The inner loop will stop when it reaches one of the termination criteria, which is defined as

   (a) Terminate after a fixed number of iterations (the maximum iteration I_{max})

   (b) After a certain number of iterations without any improvement in the cost function, e.g. all decreases are smaller than the tolerance ϵ

   (c) When the objective function reaches a pre-specified value

6. End: Complete CHO algorithm

4.9 Simulation Results and Discussion

4.9.1 PSO Dynamics Investigation

4.9.1.1 Benchmark Problems To demonstrate the accuracy of the five cases in Section 4.6.2, we simulate two nonlinear problems [46]. As Figure 4.19 shows, it

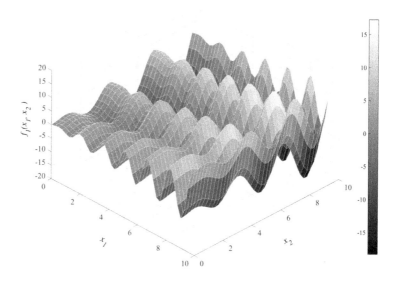

Figure 4.19   Distribution of Function f_1 in the Range [0,10]

is the distribution of the first problem $f_1 = x_1 \sin(4x_1) + 1.1x_2 \sin(2x_2)$ in the range between 0 and 10. There are several local minimum locations near the point (9.0390, 8.6682) of global minimum value, -18.5547.

Figures 4.20 and 4.21 show the distributions of the second problem (Rosenbrock) f_2 = $100(x_2 - x_1^2)^2 + (x_1 - 1)^2$ in the ranges $[-10,10]$ and $[-1,1]$, respectively (note the axes). The global minimum value is zero and locates at (1,1). It is very difficult for many optimization methods to find this point.

4.9.1.2 Selection of Parameters   According to Step 1 of the procedure in Section 4.6.1, the selection of some initial parameters is very important before the steps 2–5 are performed. Therefore, the three parameters, including maximum number of iterations, swarm size, the limited maximum velocity (V_{max}), and the radius of spline's support domains r_i, should be selected first before simulating the five techniques in Section 4.6.1. From our simulation results, the first parameter, and maximum number of iterations, are chosen as 50 for both problems; the forth parameter, r_i, is selected between 8.5 and 15.0 for both problems and we utilize $r_i = 10$ for all cases.

For the two simulation problems, there are 24 different cases with technique A or B and different random distribution (Uni, Nor1, Nor2a, Nor2b, Nor2c, or Nor2d). For example, Figure 4.22 is the distribution of the first problem fitness value f_1 with technique A and uniform distribution under different swarm sizes and maximum velocities; Figure 4.23 is the distribution of Rosenbrock problem fitness value f_2 with technique B and normal distribution N(0,1/3) under different swarm sizes and maximum velocities. Considering the effect of statistics, each simulation

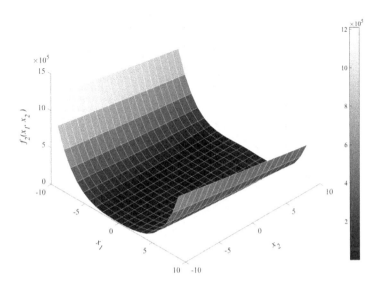

Figure 4.20   Distribution of Function f_2 in the Range $[-10,10]$

results from the average of 30 times. Hence, we obtain 24 figures like Figures 4.22 and 4.23, and Table 4.2 accounts for the selected parameters, maximum velocity (V_{max}) and swarm size, for each case. Then Table 4.3 is the selection of parameters according to the union of Table 4.2.

Notice that swarm size and maximum velocity are selected as 60 and 3.0 for Nor1; swarm size and maximum velocity are respectively selected as 40 and 1.0 for the remaining cases (Uni, Nor2a, Nor2b, Nor2c, and Nor2d). The parameters utilized in techniques C, D, and E, are the same to those in technique A.

4.9.1.3 Simulations   To consider statistics situations, after simulating 30 times, Tables 4.4 and 4.5 express the results for problem 1 and 2 with five techniques and six different random distributions, respectively. x_1^*, x_2^*, and f^* are the best solutions of the first and second variables and fitness value from 30 times, respectively; $\bar{x}_1$, $\bar{x}_2$, and $\bar{f}$ are the averages of the first and second variables and fitness value from 30 times; σ_{x_1}, σ_{x_2}, and σ_f are the standard variances of the first and second variables and fitness value after the 30 simulations.

Mendes et al. [56] and Secrest and Lamont [57] used a normal distribution with the mean value as zero, for example, Nor1 in this work. From (4.6.8), Nor1, Nor2a, Nor2b, Nor2c, and Nor2d have 50%, 30.85%, 15.87%, 2.28%, and 0.13% negative weights, respectively, as shown in Figure 4.14.

Observing the data in Tables 4.4 and 4.5, case Nor2c (normal distribution with particles obtaining 2.28% negative weights) and Nor 2d (normal distribution with

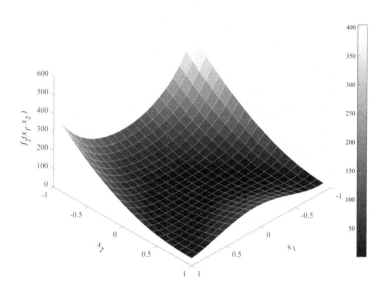

Figure 4.21   Distribution of Function f_2 in the Range $[-1,1]$

particles obtaining 0.13% negative weights) produce consistent results in terms of optimal value and smallest variance in the result. Comparing these two cases to the Uni case (uniform distribution between zero and one, therefore particles have strictly positive weighting), we conclude that a small amount of negative velocity is useful and is producing better results for these two cases. In other words, during the procedure, most particles are looking for the same searching directions (positive weights), but a few particles are looking for the different searching spaces (negative weights).

From the results of the two simulation problems, techniques A–E get agreeable performance, but it appears that E is the most costly. Considering the relative reliability and effectiveness, we will study these techniques (A–E) on a large range of problems and CPU time. In addition, after some parameters are selected appropriately, particles using normal distributions with some negative weights can indeed get better results. Therefore, we will investigate whatever of the "optimal" swarm size (40) depends on the dimension of the problem and develop an adaptive algorithm where the parameters are self-tuned as the algorithm proceeds.

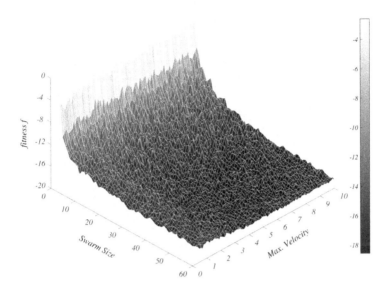

Figure 4.22   Distribution of Function f_1 with Technique A and Uni under Different Swarm Sizes and Maximum Velocities

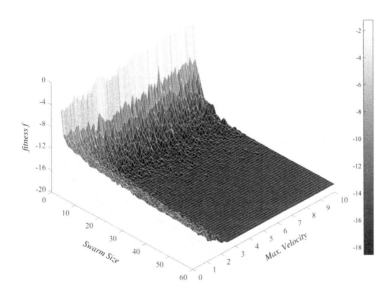

Figure 4.23   Distribution of Function f_2 with Technique B and Nor2d under Different Swarm Sizes and Maximum Velocities

Table 4.2   Selected V_{max} and Swarm Size for Each Case

		Problem 1		Problem 2	
Technique		V_{max}	Swarm Size	V_{max}	Swarm Size
A	Uni	[3.0,10.0]	35+	[0.5,3.0]	40+
	Nor1	[1.0,10.0]	25+	[2.5,4.0]	60+
	Nor2a	[0.5,3.0] or [8.5,9.2]	25+	[0.5,3.0]	40+
	Nor2b	[0.5,10.0]	25+	[0.5,3.0]	30+
	Nor2c	[2.0,10.0]	20+	[0.2,3.5]	20+
	Nor2d	[2.2,10.0]	20+	[0.2,4.0]	20+
B	Uni	[2.0,10.0]	35+	[0.5,3.5]	30+
	Nor1	[0.5,10.0]	20+	[2.5,4.0]	50+
	Nor2a	[0.5,3.2] or [8.5,9.2]	25+	[0.7,2.5]	35+
	Nor2b	[0.5,10.0]	25+	[0.4,3.5]	25+
	Nor2c	[2.0,10.0]	20+	[0.2,4.0]	20+
	Nor2d	[2.2,10.0]	20+	[0.2,4.0]	15+

Table 4.3   Selection of Parameters for Each Problem

	Problem 1	Problem 2
Maximum iteration	50	50
Swarm size	40	40 or 60*
Maximum velocity (V_{max})	3.0	1.0 or 3.0*
Spline domain radii r_i	10.0	10.0
Statistics times	30	30
Fitness function f	$f_1 = x_1 \sin(4x_1)$ $+1.1x_2 \sin(2x_2)$	$f_2 = 100(x_2 - x_1^2)^2$ $+(x_1 - 1)^2$
Range	[0, 10]	[−10, 10]
Variable optimal value	(9.0390, 8.6682)	(1, 1)
Fitness optimal value	−18.5547	0

*Swarm size and maximum velocity are chosen as 60 and 3.0 only for Nor1.

Table 4.4   Data from Each Technique and Random Variable for Problem One

Technique		$x_1 \approx 9.0390$			$x_2 \approx 8.6682$			$f \approx -18.5547$		
		x_1^*	$\bar{x}_1$	σ_{x_1}	x_2^*	$\bar{x}_2$	σ_{x_2}	f^*	$\bar{f}$	σ_f
A	Uni	9.0390	8.7775	0.5236	8.6682	7.6290	3.5648	−18.5547	−17.1456	4.2550
	Nor1	9.0505	8.9656	0.0768	8.6427	8.6764	0.0109	−18.5326	−18.1607	0.1523
	Nor2a	9.0758	8.9690	0.0954	8.6898	8.6335	0.0232	−18.4475	−17.8097	0.4676
	Nor2b	9.0336	9.0138	0.0013	8.6267	8.6550	0.0080	−18.5198	−18.2667	0.0560
	Nor2c	9.0408	9.0401	0.0005	8.6726	8.6716	0.0018	−18.5541	−18.4832	0.0048
	Nor2d	9.0364	9.0400	0.0003	8.6711	8.6689	0.0019	−18.5541	−18.4960	0.0048
B	Uni	9.0390	8.6206	0.8377	8.6682	8.4598	0.6288	−18.5547	−17.9063	2.1507
	Nor1	9.0350	8.9373	0.1639	8.6771	8.6576	0.0051	−18.5520	−18.1867	0.2571
	Nor2a	9.0295	9.0107	0.0016	8.6392	8.7065	0.0191	−18.5322	−18.0114	0.3496
	Nor2b	9.0490	9.0465	0.0022	8.6741	8.6756	0.0064	−18.5467	−18.2815	0.1179
	Nor2c	9.0408	9.0401	0.0001	8.6659	8.6716	0.0008	−18.5544	−18.5312	0.0008
	Nor2d	9.0410	9.0397	0.0001	8.6689	8.6613	0.0008	−18.5544	−18.5292	0.0005
C	Uni	9.0390	8.8299	0.8030	8.6682	8.2515	1.1677	−18.5547	−17.8859	2.0254
	Nor1	9.0340	9.0326	0.0021	8.6829	8.7025	0.0096	−18.5488	−18.2050	0.1072
	Nor2a	9.0629	8.8659	0.3732	8.6395	8.6902	0.0178	−18.4975	−17.8964	0.8067
	Nor2b	9.0292	9.0258	0.0019	8.6896	8.6782	0.0048	−18.5390	−18.3223	0.0620
	Nor2c	9.0416	9.0499	0.0007	8.6747	8.6671	0.0008	−18.5534	−18.4793	0.0401
	Nor2d	9.0399	8.9372	0.3513	8.6699	8.6965	0.0104	−18.5546	−18.0386	1.1760
D	Uni	9.0218	7.6148	4.2615	8.6721	7.9554	1.7334	−18.5331	−15.8799	5.2799
	Nor1	9.0338	8.7265	0.5881	8.6827	8.6631	0.0160	−18.5487	−17.7020	1.0070
	Nor2a	9.0364	8.9671	0.0827	8.6644	8.6386	0.0083	−18.5540	−18.2183	0.1292
	Nor2b	9.0383	9.0364	0.0002	8.6628	8.6753	0.0010	−18.5541	−18.5184	0.0014
	Nor2c	9.0348	9.0381	0.0003	8.6643	8.6711	0.0016	−18.5532	−18.5007	0.0260
	Nor2d	9.0422	9.0326	0.0007	8.6739	8.6806	0.0011	−18.5534	−18.4839	0.0170
E	Uni	9.0151	7.8783	3.0638	8.5734	7.4146	3.2577	−18.3432	−15.6750	5.2430
	Nor1	9.0422	8.8912	0.2280	8.6750	8.5826	0.3560	−18.5531	−17.9214	0.6917
	Nor2a	9.0340	8.8744	0.2338	8.6697	8.6948	0.0132	−18.5529	−17.9522	0.3418
	Nor2b	9.0410	9.0360	0.0004	8.6747	8.6799	0.0018	−18.5536	−18.4921	0.0058
	Nor2c	9.0405	9.0419	0.0001	8.6704	8.6604	0.0006	−18.5545	−18.5309	0.0006
	Nor2d	9.0387	9.0302	0.0017	8.6664	8.6610	0.0066	−18.5547	−18.3128	0.2683

Table 4.5 Data from Each Technique and Random Variable for Problem Two

Technique		x_1^*	$\bar{x}_1$	σ_{x_1}	x_2^*	$\bar{x}_2$	σ_{x_2}	f^*	$\bar{f}$	σ_f
		$x_1 \approx 9.0390$			$x_2 \approx 8.6682$			$f \approx -18.5547$		
A	Uni	1.0000	1.0500	0.0300	1.0000	1.1316	0.2160	0.	0.0315	0.0140
	Nor1	1.0000	0.9461	0.0869	1.0000	0.9833	0.2553	0.	0.1582	0.0626
	Nor2a	1.0000	1.0192	0.0099	1.0000	1.0438	0.0380	0.	0.0263	0.0013
	Nor2b	1.0000	1.0113	0.0030	1.0000	1.0267	0.0129	0.	0.0068	0.
	Nor2c	1.0000	1.0006	0.0023	1.0000	1.0042	0.0093	0.	0.0035	0.
	Nor2d	1.0014	1.0005	0.0015	1.0021	1.0032	0.0058	5.1235E−05	0.0036	0.
B	Uni	1.0001	1.0169	0.0163	1.0001	1.0498	0.1040	1.0102E−06	0.0161	0.0052
	Nor1	1.0415	1.0669	0.0735	1.0849	1.2070	0.4300	1.7254E−03	0.1368	0.0438
	Nor2a	1.0006	0.9890	0.0298	1.0000	1.0059	0.1036	1.4445E−04	0.0579	0.0063
	Nor2b	1.0000	1.0014	0.0049	1.0000	1.0091	0.0189	0.	0.0097	0.0002
	Nor2c	1.0004	1.0060	0.0029	1.0011	1.0137	0.0109	9.1504E−06	0.0039	0.
	Nor2d	1.0040	1.0151	0.0017	1.0080	1.0319	0.0071	1.6026E−05	0.0030	0.
C	Uni	1.0000	1.1024	0.0578	1.0000	1.2708	0.3871	0.	0.0693	0.0221
	Nor1	0.9820	0.9399	0.0891	0.9648	0.9728	0.3503	3.4666E−04	0.1786	0.0403
	Nor2a	1.0000	0.9695	0.0179	1.0000	0.9603	0.0553	0.	0.0554	0.0072
	Nor2b	1.0000	1.0055	0.0084	1.0000	1.0221	0.0348	0.	0.0175	0.0006
	Nor2c	1.0000	0.9879	0.0045	0.9983	0.9778	0.0172	2.8900E−04	0.0110	0.0002
	Nor2d	0.9957	1.0286	0.0113	0.9919	1.0689	0.0535	4.1675E−05	0.0221	0.0013
D	Uni	0.9762	0.7787	0.2269	0.9509	0.8202	0.1250	9.9346E−04	0.3137	0.7457
	Nor1	1.0000	0.9198	0.0282	1.0000	0.8737	0.0757	0.	0.0448	0.0072
	Nor2a	1.0000	0.9515	0.0165	1.0000	0.9216	0.0604	0.	0.0306	0.0009
	Nor2b	1.0085	0.9697	0.0053	1.0162	0.9479	0.0186	1.4833E−04	0.0094	0.0002
	Nor2c	1.0000	0.9749	0.0059	1.0000	0.9551	0.0188	0.	0.0077	0.0004
	Nor2d	0.9998	0.9706	0.0058	0.9996	0.9493	0.0217	4.0000E−08	0.0097	0.0001
E	Uni	0.9827	0.7758	0.2209	0.9670	0.8142	0.2883	4.6847E−04	0.3126	0.3966
	Nor1	0.9993	0.9908	0.0485	0.9975	1.0281	0.2506	1.2160E−04	0.0602	0.0130
	Nor2a	1.0139	1.0049	0.0253	1.0293	1.0387	0.0940	3.6398E−04	0.0471	0.0023
	Nor2b	1.0058	0.9973	0.0031	1.0099	0.9967	0.0122	3.3419E−04	0.0061	0.
	Nor2c	0.9974	0.9927	0.0027	0.9947	0.9885	0.0105	7.8998E−06	0.0045	0.
	Nor2d	1.0020	0.9910	0.0060	1.0023	0.9870	0.0211	2.9436E−04	0.0103	0.0002

4.9.2 APSO to Multiple Dimensional Problems

Two examples, Sphere and Rosenbrock problems, are utilized to demonstrate the reliability of the proposed APSO [50]. Table 4.6 expresses the selection of some parameters for each problem. For APSO, if the velocity values $V_i^j(t) < V_{min}$, then $V_i^j(t) = V_{min}$. V_{min} is the minimum velocity.

Table 4.6   Selection of Parameters for Each Problem

	Sphere	Rosenbrock
Maximum iteration	200	200
Swarm size	40	40
Distribution	Uniform	Uniform
Tolerance ϵ	10^{-8}	10^{-8}
Minimum velocity	0.5	0.5
Statistics times	15	15
Fitness function f	$f_1 = \displaystyle\sum_{i=1}^{n} x_i^2$	$f_2 = \displaystyle\sum_{i=1}^{n} [100(x_{i+1} - x_i^2)^2 + (x_i - 1)^2]$
Search range on x_i $(i = 1, \cdots, n)$	$[-10, 10]$	$[-10, 10]$

Figures 4.24, 4.25, and 4.26 show the errors of APSO and PSO with the change and unchanged updating velocity direction for sphere problem. Figure 4.26 depicts the errors of APSO and PSO with +/− standard deviation for sphere problem. Similarly, Figures 4.27, 4.28, and 4.29 are the simulations for the Rosenbrock problem, respectively.

Based on these simulation results, the proposed APSO with changed updating velocity direction obtains improved results than the generic PSO. In addition, to compare the standard deviation (Figures 4.26 and 4.29), APSO with changed updating velocity direction also yields more stable results than the generic PSO. Because the APSO technique is attempting to use secant plane information to help determine good search directions, and for very smooth problems, possessing nicely defined minima, one would expect better performance.

During the exploration stage, a maximum velocity indicates that the particles are located in an un-interesting area of the search field. Changing the velocity magnitude and its direction helps the particles to escape this area faster than changing only the magnitude of velocity. On the other hand, during the intensification stage, a maximum velocity is less likely to occur and hence the change in direction by using the velocity component maximum is not employed. A future strategy might be to use "learning rules" based on what the swarm sees in its progress to weight the influence of the adaptive secant information properly. This will be investigated in future work.

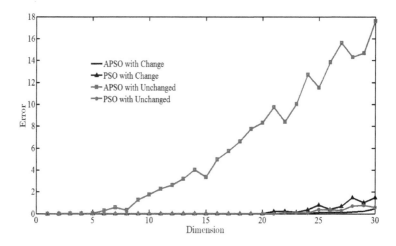

Figure 4.24   Errors of APSO and PSO for Sphere Problem

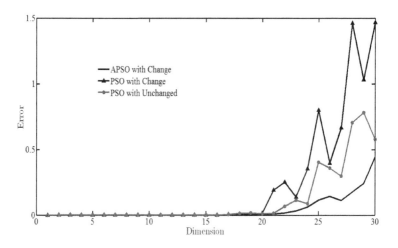

Figure 4.25   Errors of APSO and PSO for Sphere Problem

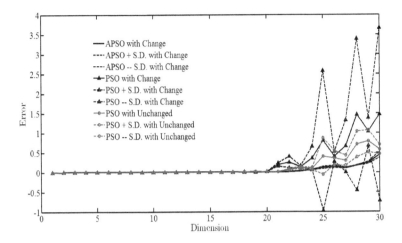

Figure 4.26   Errors of APSO and PSO with Standard Deviation for Sphere Problem

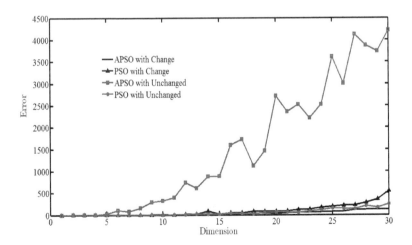

Figure 4.27   Errors of APSO and PSO for Rosenbrock Problem

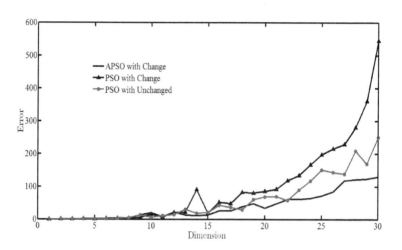

Figure 4.28   Errors of APSO and PSO for Rosenbrock Problem

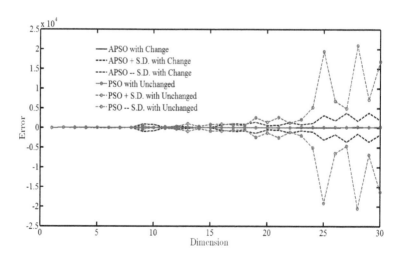

Figure 4.29   Errors of APSO and PSO with Standard Deviation for Rosenbrock Problem

4.9.3 PSO in Other Biomedical Applications

Inflammation is a key process in acute and chronic diseases. When bacteria or viruses invade the human body, the immune system is activated. One result of this activation causes leukocytes to leave the blood stream through the endothelial barrier of blood vessel walls so they can attack the microorganisms [58]. Cell surface adhesion molecules (CAMs) play a fundamental role in this process. Adhesion molecules on the surface of eukaryotic cells allow cells to specifically interact with each other and with the extracellular matrix. Four families of CAMs mediate the majority of adhesive interactions: integrins, cadherins, immunoglobulin superfamily members, and selectins [59].

Among CAMs, selectins play a key role in leukocyte extravasation. Selectins are classified by three different subsets: L-selectin (in leukocytes), E-selectin (in vascular endothelium), and P-selectin (in platelets or endothelial cells) [60]. L-selectin is expressed on most leukocyte subpopulations and is responsible for amplifying the inflammatory response through leukocyte–leukocyte interactions. E-selectin is regulated through transcription and induced in response to inflammatory stimuli. P-selectin is compartmentalized intracellularly, translocates to the cell surface early after activation, and plays an essential role in the initial recruitment of leukocytes to the site of injury during inflammation. These selectins do no act independently, but collectively contribute to inflammation. For example, when a white blood cell attaches to the wall of a blood vessel, all three of these selectins may play a role in pulling that cell out of the blood stream. While much is known about the respective properties of each individual selectin, the integrated response of combinations of selectins on tethering and capturing leukocytes remains unknown. PSO provides an opportunity to use what is known about individual rupture properties of selectins (with their respective ligands) and integrate them into a collective scenario for generating novel and testable hypotheses concerning the regulation of inflammation. Distinct kinetic and mechanical properties determine the interactions of selectin–leukocyte [60–62]. The Bell model parameters, the unstressed off-rate, and the reactive compliance were first established by a least square approximation to the linear region of a graph of rupture force against the logarithm of loading rate [63]. Hence, utilizing ruptured force to capture most molecules is significant.

4.9.3.1 Leukocyte Adhesion Molecules Modeling The average rupture forces increase linearly as a function of the natural logarithm of the loading rate. Bell [63] proposed this behavior first. In Bell's mathematical model [62], the mean rupture force F_{rup} is expressed as

$$F_{rup} = \frac{k_B T}{x_\beta} \ln \left(\frac{x_\beta}{k_{off}^0 k_B T} \right) + \frac{k_B T}{x_\beta} \ln (r_f), \qquad (4.9.1)$$

where k_B is Boltzmann constant; T is absolute temperature; x_β represents reactive compliance or mechanical bonding length; k_{off}^0 means (unstressed) dissociation rate in the absence of a pulling force; r_f is loading rate. The Bell model parameters, x_β and k_{off}^0, depict the mechanical properties of CAMs interactions. Therefore, the corresponding probability density distributions for the failure of a single pair of CAMs are calculated by

$$
P\left(F_{rup}\right) \ = \ k_{off}^0 \exp\left(\frac{x_\beta F_{rup}}{k_B T}\right) \exp\left\{ \frac{k_{off}^0 k_B T}{x_\beta r_f} \left[1 - \exp\left(\frac{x_\beta F_{rup}}{k_B T}\right)\right] \right\}.
$$
$$(4.9.2)$$

Based on (4.9.1) and (4.9.2), rupture force $F_{rup,i}$ and probability P_i of each pair i (i = 1–9) are calculated. The selection of parameters is that the Boltzmann constant k_B = 1.38065×10^{-23} (J/K); the absolute temperature T = 300 (K); the loading rate r_f = 2000 (pN/sec); all Bell parameters of receptor–ligand pairs are selected as shown in Table 4.7. Therefore, the total probability P_{tot} is computed as

$$
P_{tot}\left(F_{rup}\right) \ = \ \sum_{i=1}^{9} P_i\left(F_{rup,i}\right).
$$
$$(4.9.3)$$

Table 4.7 Bell Parameters of Each Receptor–Ligand Pairs

#	Receptor-Ligand Pairs (Ref.)	x_β (Å)	Avg.	k_{off}^0 (sec^{-1})	Avg.
1	L-selectin-PSGL-1 [64]	0.16	0.835	8.6	4.715
	L-selectin-PSGL-1 [60]	1.51		0.83	
	L-selectin mutant-PSGL-1 [64]	0.15		12.7	
2	L-selectin mutant-PSGL-1 [64]	0.12	0.127	17.3	16.1
	L-selectin mutant-PSGL-1 [64]	0.11		18.3	
3	L-selectin-neutrophil [65]	0.24	0.675	7.0	4.9
	L-selectin-neutrophil [66]	1.11		2.8	
4	E-selectin-PSGL-1 [60]	1.11	1.11	0.24	0.24
5	E-selectin-neutrophil [66]	0.18	0.245	2.6	1.65
	E-selectin-neutrophil [67]	0.31		0.7	
6	P-selectin-LS174T [61]	0.9	0.9	2.96	2.96
	P-selectin-PSGL-1 [60]	1.35		0.18	
7	P-selectin-PSGL-1 [64]	0.29	1.38	1.1	0.434
	P-selectin-PSGL-1 [68]	2.5		0.022	
	P-selectin mutant-PSGL-1 [64]	0.24		1.8	
8	P-selectin mutant-PSGL-1 [64]	0.33	0.33	1.7	1.7
	P-selectin mutant-PSGL-1 [64]	0.42		1.6	
9	P-selectin-neutrophil [66]	0.39	0.395	2.4	1.665
	P-selectin-neutrophil [69]	0.40		0.93	

Figure 4.30 shows the probability of counts of events under rupture force for each pair. The result of the PSO algorithm displays the optimal rupture force as F^*_{rup} = 141.2424 (pN) with a maximum probability P^*_{tot} = 88.4189.

These results demonstrate the utility of PSO in generating predictions about the integrated effects of multiple selectin–ligand pairs. These predictions are then used to generate testable hypotheses. Use of this system will speed the advancement of understanding how expression and regulation of multiple selectins–ligands contributes to inflammation *in vivo* [31, 70].

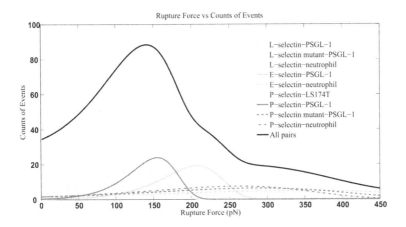

Figure 4.30   Rupture Force versus Counts of Events for Different Pairs

4.9.4   CHO to Multiple Dimensional Problems

Two examples (Hyperbolic and Rosenbrock problems) are utilized to demonstrate the reliability of the proposed CHO. The parameter selection of TS includes the length of the tabu list N_1 = 6, the length of the promising list N_2 = 6, the initial radius of neighbor r_1 = 0.25, the radius of tabu balls r_2 = 0.125, the radius of promising balls r_3 = 0.06, and the maximum number of iterations I_{max} = 200. The parameter selection of GA includes the population size of generation = 48, the number of chromosomes kept for mating = 12, the mutation rate = 0.04, and the maximum number of iterations = 200. As for PSO, the parameters include the swarm size = 40, the maximum velocity = 2.5, uniform distribution is chosen as random variables, the tolerance $\epsilon = 10^{-8}$, and the maximum number of iterations = 200. The proposed CHO utilizes the same parameter selections for TS and PSO and the constant radius of the promising area r_b = 0.5 [22].

The Hyperbolic and Rosenbrock cost functions are, respectively, described as

$$f_{Hyperbolic} = \sum_{i=1}^{n} x_i^2, \tag{4.9.4}$$

$$f_{Rosenbrock} = \sum_{i=1}^{n} \left[100(x_{i+1} - x_i^2)^2 + (x_i - 1)^2 \right]. \tag{4.9.5}$$

Tables 4.8 and 4.9 show the mean and standard deviation errors of five algorithms (ECTS, CGA, PSO, ECTS combined with CGA, and CHO) for Hyperbolic and Rosenbrock problems, respectively, after 30 simulations. The search ranges of both problems are on $[-10,10]$. Based on the simulation results, PSO can obtain an ex-

Table 4.8 Mean and Standard Deviation Errors of Five Algorithms for Hyperbolic Problem

n	ECTS	CGA	PSO	ECTS + CGA	CHO
2	0.0179 ± 0.0218	0.0002 ± 0.0007	$3.1 \times 10^{-8} \pm 3.8 \times 10^{-8}$	$0.0001 \pm 3.9 \times 10^{-5}$	$1.5 \times 10^{-9} \pm 1.5 \times 10^{-9}$
3	0.0190 ± 0.0146	0.0028 ± 0.0041	$1.2 \times 10^{-6} \pm 8.0 \times 10^{-7}$	0.0026 ± 0.0061	$1.1 \times 10^{-7} \pm 2.1 \times 10^{-7}$
5	0.0238 ± 0.0261	0.0471 ± 0.0265	0.0001 ± 0.0001	0.0207 ± 0.0260	$4.7 \times 10^{-6} \pm 1.0 \times 10^{-5}$
10	0.0303 ± 0.0327	0.6255 ± 0.3012	0.0011 ± 0.0014	0.0342 ± 0.0285	0.0003 ± 0.0003
20	0.0501 ± 0.0506	3.8993 ± 0.8040	0.2263 ± 0.1611	0.0432 ± 0.0400	0.0297 ± 0.0903
30	0.0649 ± 0.0782	8.1319 ± 1.1536	2.0799 ± 1.0858	0.1035 ± 0.0760	0.2698 ± 0.6996
50	0.1339 ± 0.1240	16.1665 ± 1.4039	10.8841 ± 2.2802	0.1273 ± 0.1276	0.4632 ± 2.0628
100	0.2383 ± 0.2189	38.0997 ± 1.8511	39.8710 ± 3.9074	0.2073 ± 0.2442	1.2373 ± 5.9897
300	0.6314 ± 0.6334	129.7538 ± 3.0646	176.5882 ± 9.2199	0.8907 ± 0.9396	0.4526 ± 0.3736
500	1.1307 ± 1.0052	222.8210 ± 5.9827	318.8833 ± 11.5058	1.0688 ± 0.9051	1.0291 ± 0.9075

Table 4.9 Mean and Standard Deviation Errors of Five Algorithms for Rosenbrock Problem

n	ECTS	CGA	PSO	ECTS + CGA	CHO
2	0.0137 ± 0.0138	0.5242 ± 0.2789	0.0003 ± 0.0004	0.2703 ± 0.7590	0.0336 ± 0.1327
3	0.0148 ± 0.0151	1.0620 ± 0.3752	0.6215 ± 1.0571	0.1653 ± 0.0796	0.0891 ± 0.1511
5	0.0195 ± 0.0203	2.1632 ± 0.9059	2.2764 ± 0.9663	0.1326 ± 0.0486	0.1561 ± 0.2368
10	0.0373 ± 0.0397	4.2154 ± 2.5281	7.9283 ± 1.8828	0.0619 ± 0.0279	0.5464 ± 1.8083
20	0.0568 ± 0.0568	8.0005 ± 3.1585	16.4347 ± 3.4651	0.0535 ± 0.0209	0.0893 ± 0.0486
30	0.0649 ± 0.0578	12.4630 ± 3.9880	23.0807 ± 3.5200	0.0720 ± 0.0315	2.4114 ± 6.9677
50	0.1233 ± 0.1500	19.8758 ± 4.3163	39.4033 ± 4.3694	0.0984 ± 0.0436	1.7678 ± 8.7988
100	0.2333 ± 0.1938	43.0156 ± 4.9253	88.2320 ± 6.8456	0.1769 ± 0.1233	0.2243 ± 0.0775
300	0.4577 ± 0.4289	145.6715 ± 7.2527	304.8118 ± 15.5815	0.5147 ± 0.4542	3.6957 ± 10.7818
500	1.4962 ± 1.8032	246.2241 ± 8.9461	531.4223 ± 14.6164	0.8567 ± 0.8093	1.0540 ± 0.8991

cellent performance on less than 10-dimensional Hyperbolic problems. However, PSO cannot find the global minimum on larger than 50-dimensional Hyperbolic problems. Similarly, GA can get a good performance on less than 10-dimensional Hyperbolic problems, but GA does not perform as well as PSO. In addition, GA cannot achieve the global minimum for larger than 30-dimensional Hyperbolic problems. TS looks very robust for all dimensional Hyperbolic problems, but PSO is much better than TS on less than 10-dimensional Hyperbolic problems. Both hybrid algorithms, ECTS+CGA and CHO, combine the benefits of TS and GA or PSO on the entire range of Hyperbolic problems. Furthermore, CHO is more robust than ECTS+GA on less than 20-dimensional Hyperbolic problems.

As for Rosenbrock problems, the simulation results show that the similar results as on Hyperbolic problems are obtained. However, PSO and GA have more difficulty in finding the global minimum, especially on the higher-dimensional Rosenbrock problems. TS still retains robust characteristics. Similarly, both hybrid algorithms, ECTS+CGA and CHO, combine the benefits of TS and GA or PSO on all dimensional Rosenbrock problems, except for the 300-dimensional Rosenbrock problem. On the higher-dimensional Rosenbrock problems, CHO shows a very sensitive dependence on selecting the most promising area from the promising list.

This work shows the comparison of five algorithms (ECTS, CGA, PSO, ECTS combined with CGA, and CHO) on multi-dimensional Hyperbolic and Rosenbrock problems. The simulation results show that the proposed CHO algorithm combines the advantages of TS and PSO and obtains robust results. However, on the higher-dimensional problems, CHO shows a sensitive dependence on selecting the most promising area from the promising list. The constant radius (r_b) of the chosen promising area plays a key role. Therefore, this parameter selection is studied in future work.

Bibliography

[1] F. O. Karray and C. De Silva. *Soft Computing and Intelligent Systems Design: Theory, Tools and Applications*. Pearson Educational Limited, Harlow, UK, 2004.

[2] A. Konar. *Computational Intelligence: Principles, Techniques and Applications*. Springer-Verlag, Berlin, Germany, 2005.

[3] R. R. Seeley, T. D. Stephens, and P. Tate. *Anatomy and Physiology, Eighth Edition*. The McGraw-Hill, New York, USA, 2007.

[4] J.-S. R. Jang, C.-T. Sun, and E. Mizutani. *Neuro-Fuzzy and Soft Computing: A Computational Approach to Learning and Machine Intelligence*. Prentice Hall PTR, Upper Saddle River, New Jersey, USA, 1997.

[5] H. T. Nguyen, N. R. Prasad, C. L. Walker, and E. A. Walker. *A First Course in Fuzzy and Neural Control*. Chapman & Hall/CRC, Boca Raton, Florida, USA, 2003.

[6] F. L. Lewis, D. M. Dawson, and C. T. Abdallah. *Robot Manipulators Control: Second Edition, Revised and Expanded*. Marcel Dekker, Inc., New York, USA, 2004.

[7] L. A. Zadeh. Fuzzy sets. *Information and Control*, 8:338–353, 1965.

[8] S. Sumathi and Surekha Paneerselvam. *Computational Intelligence Paradigms: Theory and Applications Using Matlab*. CRC Press, Boca Raton, Florida, USA, 2010.

[9] E. H. Mamdani and S. Assilian. An experiment in linguistic synthesis with a fuzzy logic controller. *International Journal of Man–Machine Studies*, 7(1):1–13, 1975.

[10] T. Takagi and M. Sugeno. Fuzzy identification of systems and its application to modeling and control. *IEEE Transactions on Systems, Man, and Cybernetics*, 15(1):116–132, 1985.

[11] F. L. Lewis, S. Jagannathan, and A. Yesildirek. *Neural Network Control of Robotic Manipulators and Nonlinear Systems*. Taylor & Francis, London, UK, 1999.

[12] B. Widrow and M. E. Hoff. Adaptive switching circuits. *IRE WESCON Convention Record*, pp. 96–104, 1960.

[13] H. Takagi. Fusion technology of neural networks and fuzzy systems: A chronicled progression from the laboratory to our daily lives. *International Journal of Applied Mathematics and Computer Science*, 10(4):647–673, 2000.

[14] J.-S. R. Jang. ANFIS: Adaptive-network-based fuzzy inference system. *IEEE Transactions on Systems, Man, and Cybernetics*, 23(665–685):3–12, 1993.

[15] C.-H. Chen, D. S. Naidu, and M. P. Schoen. Adaptive control for a five-fingered prosthetic hand with unknown mass and inertia. *World Scientific and Engineering Academy and Society (WSEAS) Journal on Systems*, 10(5):148–161, May 2011.

[16] C.-H. Chen and D. S. Naidu. "Fusion of fuzzy logic and PD control for a five-fingered smart prosthetic hand," in *Proceedings of the 2011 IEEE International Conference on Fuzzy Systems (FUZZ–IEEE 2011)*, pp. 2108–2115, Taipei, Taiwan, June 27–30, 2011.

[17] C.-H. Chen and D. S. Naidu. Hybrid control strategies for a five-finger robotic hand. *Biomedical Signal Processing and Control*, 8(4):382–390, July 2013.

[18] F. Glover. Tabu search-part I. *ORSA Journal of Computing*, 10(3):190–205, 1989.

[19] F. Glover. Tabu search-part II. *ORSA Journal of Computing*, 2(1):4–32, 1990.

[20] P. Siarry and G. Berthiau. Fitting tabu search to optimize functions of continuous variables. *International Journal for Numerical Methods in Engineering*, 40:2449–2459, 1997.

[21] R. Chelouah and P. Siarry. Tabu search applied to global optimization. *European Journal of Operational Research*, 123:256–270, 2000.

[22] C.-H. Chen, M. P. Schoen, and K. W. Bosworth. "A condensed hybrid optimization algorithm using enhanced continuous tabu search and particle swarm optimization," in *Proceedings of the ASME 2009 Dynamic Systems and Control Conference (DSCC)*, Hollywood, California, USA, October 12–14, 2009 (No. DSCC2009-2526).

[23] R. Chelouah and P. Siarry. A continuous genetic algorithm design for global optimization of multimodal functions. *Journal of Heuristics*, 6:191–213, 2000.

[24] C.-H. Chen and D. S. Naidu. "Hybrid genetic algorithm PID control for a five-fingered smart prosthetic hand," in *Proceedings of the Sixth International Conference on Circuits, Systems and Signals (CSS'11)*, pp. 57–63, Vouliagmeni Beach, Athens, Greece, March 7–9, 2012.

[25] J. Kennedy and R. Eberhart. "Particle swarm optimization," in *IEEE International Conference on Neural Networks*, 4: 1942–1948, 1995.

[26] J. Kennedy, R. Eberhart, and Y. Shi. *Swarm intelligence*. Morgan Kaufmann Publishers, San Francisco, California, USA, 2001.

[27] A. P. Engelbrecht. *Fundamentals of Computational Swarm Intelligence*. John Wiley & Sons, 2006.

[28] M. Clerc. *Particle Swarm Optimization*. ISTE Publishing Company, 2006.

[29] R. C. Eberhart and Y. Shi. *Computational Intelligence Concepts to Implementations*. Morgan Kaufmann Publishers, 2007.

[30] A. P. Engelbrecht. *Computationa Intelligence: An Introduction, Second Edition*. John Wiley & Sons, 2007.

[31] C.-H. Chen, K. W. Bosworth, M. P. Schoen, S. E. Bearden, D. S. Naidu, and A. Perez-Gracia. "A study of particle swarm optimization on leukocyte adhesion molecules and control strategies for smart prosthetic hand," in *2008 IEEE Swarm Intelligence Symposium (IEEE SIS08)*, St. Louis, Missouri, USA, September 21–23, 2008.

[32] C.-H. Chen, D. S. Naidu, A. Perez-Gracia, and M. P. Schoen. "Fusion of hard and soft control techniques for prosthetic hand," in *Proceedings of the International Association of Science and Technology for Development (IASTED) International Conference on Intelligent Systems and Control (ISC 2008)*, pp. 120–125, Orlando, Florida, USA, November 16–18, 2008.

[33] Y. Shi and R. Eberhart. "Parameter selection in particle swarm optimization," in *Evolutionary Programming VII: Proceedings of the Seventh Annual Conference on Evolutionary Programming*, pp. 591–600, New York, USA, 1998.

[34] Y. Shi and R. Eberhart. "Empirical study of particle swarm optimization," in *Proceedings of the IEEE Congress on Evolutionary Computation (IEEE Press)*, pp. 1945–1950, 1999.

[35] R. Eberhart and Y. Shi. "Comparing inertia weights and constriction factors in particle swarm optimization," in *Proceedings of the IEEE Congress on Evolutionary Computation*, pp. 84–88, San Diego, California, USA, 2000.

[36] M. Clerc and J. Kennedy. The particle swarm-explosion, stability, and convergence in a multi-dimensional complex space. *IEEE Transactions on Evolutionary Computation*, 6(1):58–73, 2002.

[37] H. J. Meng, P. Zheng, R. Y. Wu, X. J. Hao, and Z. Xie. "A hybrid particle swarm algorithm with embedded chaotic search,' in *Proceedings of the IEEE Conference on Cybernetics and Intelligent Systems*, pp. 367–371, Singapore, 2004.

[38] S. H. Ho, S. Yang, G. Ni, E. W. C. Lo, and H. C. Wong. A particle swarm optimization-based method for multiobjective design optimizations. *IEEE Transactions on Magnetics*, 41(5):1756–1759, 2005.

[39] J. H. Seo, C. H. Im, C. G. Heo, J. K. Kim, H. K. Jung, and C. G. Lee. Multimodal function optimization based on particle swarm optimization. *IEEE Transactions on Magnetics*, 42(4):1095–1098, 2006.

[40] R. A. Krohling and L. dos S. Coelho. Coevolutionary particle swarm optimization using gaussian distribution for solving constrained optimization problems. *IEEE Transactions on Systems, Man, and Cybernetics, Part B: Cybernetics*, 36(6):1407–1416, 2006.

[41] R. Brits, A. P. Engelbrecht, and F. van den Bergh. Locating multiple optima using particle swarm optimization. *Applied Mathematics and Computation*, 189(2):1859–1883, 2007.

[42] F. van den Bergh and A. P. Engelbrecht. A study of particle swarm optimization particle trajectories. *Information Sciences*, 176(8):937–971, 2006.

[43] H. Someya. "Cautious particle swarm," in *2008 IEEE Swarm Intelligence Symposium*, St. Louis, Missouri, USA, September 21–23, 2008.

[44] L.-Y. Chuang, S.-W. Tsai, and C.-H. Yang. "Catfish particle swarm optimization," in *2008 IEEE Swarm Intelligence Symposium*, St. Louis, Missouri, USA, September 21–23, 2008.

[45] F. Pan, X. Hu, R. Eberhart, and Y. Chen. "An analysis of bare bones particle swarm," in *2008 IEEE Swarm Intelligence Symposium*, St. Louis, Missouri, USA, September 21–23, 2008.

[46] C.-H. Chen, K. W. Bosworth, and M. P. Schoen. "Investigation of particle swarm optimization dynamics," in *Proceedings of International Mechanical Engineering Congress and Exposition (IMECE) 2007*, Seattle, Washington, USA, November 11–15, 2007 (No. IMECE2007-41343).

[47] R. Eberhart and Y. Shi. Guest editorial special on particle swarm optimization. *IEEE Transactions on Evolutionary Computation*, 8(3):201–203, 2004.

[48] S. N. Atluri, H. G. Kim, and J. Y. Cho. A critical assessment of the truly Meshless Local Petrov-Galerkin (MLPG), and Local Boundary Integral Equation (LBIE) methods. *Computational Mechanics*, 24:348–372, 1999.

[49] W. Pan, F. Zhun, F. Shan, and Z. Yun. "Study on a novel hybrid adaptive genetic algorithm embedding conjugate gradient algorithm," in *Proceedings of the Third World Congress on Intelligent Control and Automation*, Hefei, P. R. China, June 28–July 2, 2000.

[50] C.-H. Chen, K. W. Bosworth, and M. P. Schoen. "An adaptive particle swarm method to multiple dimensional problems," in *Proceedings of the International Association of Science and Technology for Development (IASTED) International Symposium on Computational Biology and Bioinformatics (CBB 2008)*, pp. 260–265, Orlando, Florida, USA, November 16–18, 2008.

[51] D. E. Goldberg. *Genetic Algorithm in Search, Optimization and Machine Learning*. Addison Wesley Longman, Inc., 1989.

[52] Y. Guo, L.-P. Chen, S. Wang, and J. Zhao. A new simulation optimization system for the parameters of a machine cell simulation model. *International Journal of Advanced Manufacturing Technology*, 21:620–626, 2003.

[53] L. Pladugu, M. P. Schoen, and B. Williams. "Intelligent techniques for star-pattern recognition," in *Proceedings of IMECE 2003*, Washington, District of Columbia, USA, November 16–17, 2003.

[54] B. Ramkumar, M. P. Schoen, F. Lin, and B. G. Williams. "Hybrid optimization algorithm using enhanced continuous tabu search and genetic algorithm for parameter estimation," in *Proceedings of International Mechanical Engineering Congress and Exposition (IMECE) 2006*, Chicago, Illinois, USA, November 5–10, 2006 (No. IMECE2006–13374).

[55] B. Ramkumar, M. P. Schoen, and F. Lin. "Application of an intelligent hybrid optimization technique for parameter estimation in the presence of colored noise," in *Proceedings of International Mechanical Engineering Congress and Exposition (IMECE) 2007*, Seattle, Washington, USA, November 11–15, 2007 (No. IMECE2007–41352).

[56] R. Mendes, J. Kennedy, and J. Neves. "Watch thy neighbor or how the swarm can learn from its environment," in *Proceedings of the IEEE Swarm Intelligence Symposium 2003 (SIS 2003)*, pp. 88–94, Indianapolis, Indiana, USA, 2003.

[57] B. R. Secrest and G. B. Lamont. "Visualizing particle swarm optimizationx-gaussian particle swarm optimization," in *Proceedings of the IEEE Swarm Intelligence Symposium 2003 (SIS 2003)*, pp. 198–204, Indianapolis, Indiana, USA, 2003.

[58] P. J. Delves and I. M. Roitt. Advances in immunology: The immune system. *The New England Journal of Medicine*, 343(1):37–50, July 6, 2000.

[59] R. Rhoades and R. Pflanzer. *Human Physiology*. Thomson Brooks/Cole, Fourth Edition, 2003.

[60] W. D. Hanley, D. Wirtz, and K. Konstantopoulos. Distinct kinetic and mechanical properties govern selectin-leukocyte interactions. *Journal of Cell Science*, 117(12):2503–2511, 2004.

[61] W. Hanley, O. McCarty, S. Jadhav, Y. Tseng, D. Wirtz, and K. Konstantopoulos. Single molecule characterization of p-selectin/ligand binding. *The Journal of Biological Chemistry*, 278(12):10556–10561, 2003.

[62] J. te Riet, A. W. Zimmerman, A. Cambi, B. Joosten, S. Speller, R. Torensma, F. N. van Leeuwen, C. G. Figdor, and F. de Lange. Distinct kinetic and mechanical properties govern alcam-mediated interactions as shown by single-molecule force spectroscopy. *Journal of Cell Science*, 120(22):3965–3976, 2007.

[63] G. I. Bell. Models for the specific adhesion of cells to cells. *Science*, 200:618–627, 1978.

[64] V. Ramachandran, M. U. Nollert, W.-J. Liu, H. Qiu, R. D. Cummings, C. C. Zhu, and R. P. McEver. Tyrosine replacement in p-selectin glycoprotein ligand-1 affects distinct kinetic and mechanical properties of bonds with p- and l-selectin. *Proceedings of the National Academy of Sciences of the United States of America*, 96(24):13771–13776, 1999.

[65] R. Alon, S. Chen, R. Fuhlbrigge, K. D. Puri, and T. A. Springer. The kinetics and shear threshold of transient and rolling interactions of l-selectin with its ligand on leukocytes. *Proceedings of the National Academy of Sciences of the United States of America*, 95(20):11631–11636, 1998.

[66] M. J. Smith, E. L. Berg, and M. B. Lawrence. A direct comparison of selectin-mediated transient, adhesive events using high temporal resolution. *Biophysical Journal*, 77(6):3371–3383, 1999.

[67] R. Alon, S. Chen, K. D. Puri, E. B. Finger, and T. A. Springer. The kinetics of L-selectin tethers and the mechanics of selectin-mediated rolling. *The Journal of Cell Biology*, 138(5):1169–1180, 1997.

[68] J. Fritz, A. G. Katopodis, F. Kolbinger, and D. Anselmetti. Force-mediated kinetics of single P-selectin/ligand complexes observed by atomic force microscopy. *Proceedings of the National Academy of Sciences of the United States of America*, 95(21):12283–12288, 1998.

[69] R. Alon, D. A. Hammer, and T. A. Springer. Lifetime of the P-selectin-carbohydrate bond and its response to tensile force in hydrodynamic flow. *Nature (London)*, 374(6522):539–542, 1995.

[70] C.-H. Chen. *Hybrid Control Strategies for Smart Prosthetic Hand*. PhD Dissertation, Idaho State University, Pocatello, Idaho, USA, May 2009.

CHAPTER 5

FUSION OF HARD AND SOFT CONTROL STRATEGIES I

This chapter is a continuation of Chapter 4 focusing on the fusion (or hybrid or integration) of hard and soft control strategies to the robotic/prosthetic hand. In particular, PID, optimal and adaptive hard control techniques are integrated with soft control technique of ANFIS for trajectory planning. With a brief introduction to feedback linearization in Section 5.1, we present PID control in Section 5.2, optimal control in Section 5.3, and adaptive control in Section 5.4. Simulations are given in Section 5.5.

5.1 Feedback Linearization

If the nonlinear term $N(q, \dot{q})$ in (3.3.16) only models the Coriolis/centripetal and gravity terms, then the dynamic equations of thumb and all fingers are rewritten as below.

$$M(q(t))\ddot{q}(t) + N(q(t), \dot{q}(t)) = \tau(t), \qquad (5.1.1)$$

where $N(q(t), \dot{q}(t)) = C(q(t), \dot{q}(t)) + G(q(t))$ represents nonlinear terms.

The nonlinear dynamics represented by (5.1.1) is to be converted into a linear state-variable system by finding a transformation using feedback linearization tech-

Fusion of Hard and Soft Control Strategies for the Robotic Hand, By C.-H. Chen and D. S. Naidu **161**
© 2017 by the Institute of Electrical and Electronic Engineers, Inc. Published 2017 by John Wiley & Sons, Inc.

nique [1, 2]. Alternative state-space equations of the dynamics [2, 3] are obtained by defining the angular position/velocity state $\mathbf{x}(t)$ of the joints as

$$\mathbf{x}(t) = [\mathbf{q}'(t) \quad \dot{\mathbf{q}}'(t)]'. \tag{5.1.2}$$

Let us repeat the dynamical model and rewrite (5.1.1) as

$$\frac{d}{dt}\dot{\mathbf{q}}(t) = \ddot{\mathbf{q}}(t) = -[\mathbf{M}(\mathbf{q}(t))]^{-1}\mathbf{N}(\mathbf{q}(t), \dot{\mathbf{q}}(t)) + [\mathbf{M}(\mathbf{q}(t))]^{-1}\boldsymbol{\tau}(t). \tag{5.1.3}$$

5.1.1   State Variable Representation

Form 1: Choosing the state variables (5.1.2), one way of representing the robot hand dynamics (5.1.3) for a robotic hand finger, in general, in state-space form is

$$\dot{\mathbf{x}}(t) = \begin{bmatrix} \dot{\mathbf{q}}(t) \\ \ddot{\mathbf{q}}(t) \end{bmatrix},$$

$$= \begin{bmatrix} \dot{\mathbf{q}}(t) \\ -[\mathbf{M}(\mathbf{q}(t))]^{-1}\mathbf{N}(\mathbf{q}(t), \dot{\mathbf{q}}(t)) \end{bmatrix} + \begin{bmatrix} 0 \\ [\mathbf{M}(\mathbf{q}(t))]^{-1} \end{bmatrix}\boldsymbol{\tau}(t), \tag{5.1.4}$$

which is typical of a nonlinear system of the form

$$\dot{\mathbf{x}}(t) = \mathbf{f}(\mathbf{x}(t), \mathbf{u}(t), t), \tag{5.1.5}$$

where $\mathbf{u}(t) = \boldsymbol{\tau}(t)$.

Form 2: Alternatively, we can write (3.3.16) in another general state-space form as

$$\begin{bmatrix} \mathbf{I} & 0 \\ 0 & \mathbf{M}(\mathbf{q}(t)) \end{bmatrix} \begin{bmatrix} \dot{\mathbf{q}}(t) \\ \ddot{\mathbf{q}}(t) \end{bmatrix} = \begin{bmatrix} \dot{\mathbf{q}}(t) \\ -\mathbf{N}(\mathbf{q}(t), \dot{\mathbf{q}}(t)) \end{bmatrix} + \begin{bmatrix} 0 \\ \mathbf{I} \end{bmatrix}\boldsymbol{\tau}(t). \tag{5.1.6}$$

On the other hand, from (5.1.2) and (5.1.3), another alternate *linear* state-variable equation in *Brunovsky canonical form* for (3.3.16) is written as

$$\dot{\mathbf{x}}(t) = \begin{bmatrix} 0 & \mathbf{I} \\ 0 & 0 \end{bmatrix}\mathbf{x}(t) + \begin{bmatrix} 0 \\ \mathbf{I} \end{bmatrix}\mathbf{u}(t) \tag{5.1.7}$$

with its control input vector given by

$$\mathbf{u}(t) = -[\mathbf{M}(\mathbf{q}(t))]^{-1}\mathbf{N}(\mathbf{q}(t), \dot{\mathbf{q}}(t)) + [\mathbf{M}(\mathbf{q}(t))]^{-1}\boldsymbol{\tau}(t). \tag{5.1.8}$$

Let us suppose the robotic hand is required to track the desired trajectory $\mathbf{q}_d(t)$ described under path generation or tracking. Then, the tracking error $\mathbf{e}(t)$ is defined as

$$\mathbf{e}(t) = \mathbf{q}_d(t) - \mathbf{q}(t). \tag{5.1.9}$$

Here, $\mathbf{q}_d(t)$ is the *desired* angle vector of joints and is obtained by (2.5.2), (2.3.4) and (2.3.7); $\mathbf{q}(t)$ is the *actual* angle vector of joints. Differentiating (5.1.9) twice, to get

$$\dot{\mathbf{e}}(t) \;=\; \dot{\mathbf{q}}_d(t) - \dot{\mathbf{q}}(t), \quad \ddot{\mathbf{e}}(t) = \ddot{\mathbf{q}}_d(t) - \ddot{\mathbf{q}}(t). \tag{5.1.10}$$

Substituting (5.1.2) into (5.1.10) yields

$$\ddot{\mathbf{e}}(t) \;=\; \ddot{\mathbf{q}}_d(t) + [\mathbf{M}(\mathbf{q}(t))]^{-1} [\mathbf{N}(\mathbf{q}(t), \dot{\mathbf{q}}(t)) - \boldsymbol{\tau}(t)] \tag{5.1.11}$$

from which the control function is defined as

$$\mathbf{u}(t) \;=\; \ddot{\mathbf{q}}_d(t) + [\mathbf{M}(\mathbf{q}(t))]^{-1} [\mathbf{N}(\mathbf{q}(t), \dot{\mathbf{q}}(t)) - \boldsymbol{\tau}(t)]. \tag{5.1.12}$$

This is often called the *feedback linearization* control law, which is also inverted to express it as

$$\boldsymbol{\tau}(t) \;=\; \mathbf{M}(\mathbf{q}(t)) [\ddot{\mathbf{q}}_d(t) - \mathbf{u}(t)] + \mathbf{N}(\mathbf{q}(t), \dot{\mathbf{q}}(t)). \tag{5.1.13}$$

Using the relations (5.1.10) and (5.1.12), and defining state vector $\mathbf{x}(t) = [\mathbf{e}'(t) \; \dot{\mathbf{e}}'(t)]'$, the *tracking error dynamics* is written as

$$\dot{\mathbf{x}}(t) = \begin{bmatrix} \mathbf{0} & \mathbf{I} \\ \mathbf{0} & \mathbf{0} \end{bmatrix} \mathbf{x}(t) + \begin{bmatrix} \mathbf{0} \\ \mathbf{I} \end{bmatrix} \mathbf{u}(t). \tag{5.1.14}$$

Note that this is in the form of a *linear* system such as

$$\dot{\mathbf{x}}(t) = \mathbf{A}\mathbf{x}(t) + \mathbf{B}\mathbf{u}(t) \tag{5.1.15}$$

with its control input vector given by

$$\mathbf{u}(t) = -\mathbf{M}^{-1}(\mathbf{q}(t)) [\mathbf{N}(\mathbf{q}(t), \dot{\mathbf{q}}(t)) - \boldsymbol{\tau}(t)]. \tag{5.1.16}$$

The required torque of all joints is then calculated by

$$\boldsymbol{\tau}(t) = \mathbf{M}(\mathbf{q}(t))\mathbf{u}(t) + \mathbf{N}(\mathbf{q}(t), \dot{\mathbf{q}}(t)). \tag{5.1.17}$$

5.2 PD/PI/PID Controllers

Proportional-derivative (PD) controller is the simplest closed-loop controller which is used to control the robotic manipulators. PD controller uses the feedback combination of proportional part (position) and derivative part (velocity). However, if the robotic manipulator dynamics contains the particular vectors of gravitational terms ((3.3.15) and (3.4.5)), then the position control objective cannot be reached by the simple PD control law [4]. Hence, to satisfy the position control objective, an integral component is driven into proportional integral (PI) and proportional integral derivative (PID) controllers. PD, PI and PID controllers are briefly described in this section [5].

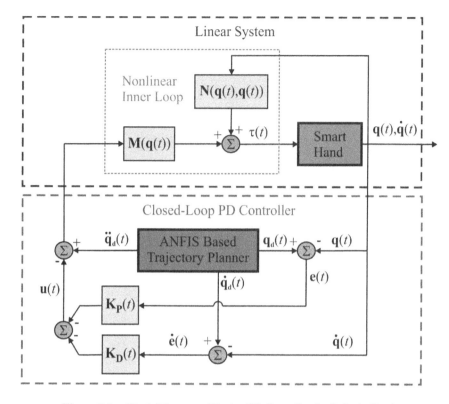

Figure 5.1   Block Diagram of Fusion PD Controller for Robotic Hand

5.2.1   PD Controller

Figure 5.1 shows the block diagram of a PD controller [2]. The control signal $\mathbf{u}(t)$ becomes

$$\mathbf{u}(t) \;=\; -\mathbf{K_P}\,\mathbf{e}(t) - \mathbf{K_D}\,\dot{\mathbf{e}}(t) \qquad (5.2.1)$$

with the proportional $\mathbf{K_P}$ and derivative $\mathbf{K_D}$ diagonal gain matrices. Then, the closed-loop error dynamics and state-space form are written as

$$\ddot{\mathbf{e}}(t) + \mathbf{K_D}\,\dot{\mathbf{e}}(t) + \mathbf{K_P}\,\mathbf{e}(t) \;=\; \mathbf{0}, \qquad (5.2.2)$$

$$\frac{d}{dt}\begin{bmatrix}\mathbf{e}(t)\\ \dot{\mathbf{e}}(t)\end{bmatrix} \;=\; \begin{bmatrix}\mathbf{0} & \mathbf{I}\\ -\mathbf{K_P} & -\mathbf{K_D}\end{bmatrix}\begin{bmatrix}\mathbf{e}(t)\\ \dot{\mathbf{e}}(t)\end{bmatrix} + \begin{bmatrix}\mathbf{0}\\ \mathbf{I}\end{bmatrix}\mathbf{u}(t) \qquad (5.2.3)$$

with

$$\boldsymbol{\tau}(t) \;=\; \mathbf{M}(\mathbf{q}(t))\,[\ddot{\mathbf{q}}_d(t) - \mathbf{u}(t)] + \mathbf{N}(\mathbf{q}(t), \dot{\mathbf{q}}(t)) \qquad (5.2.4)$$

or

$$\boldsymbol{\tau}(t) \;=\; \mathbf{M}(\mathbf{q}(t))\,[\ddot{\mathbf{q}}_d(t) + \mathbf{K_D}\,\dot{\mathbf{e}}(t) + \mathbf{K_P}\,\mathbf{e}(t)] + \mathbf{N}(\mathbf{q}(t), \dot{\mathbf{q}}(t)). \;(5.2.5)$$

5.2.2 PI Controller

Figure 5.2 shows the block diagram of a PI controller with control signal as

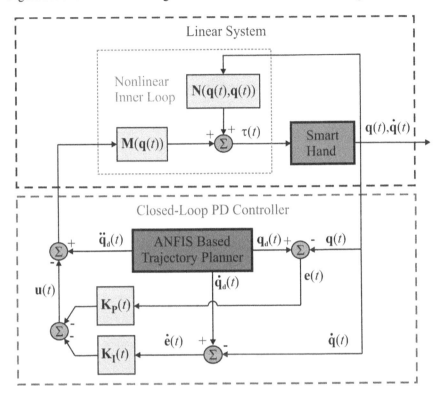

Figure 5.2   Block Diagram of Fusion PI Controller for Robotic Hand

$$\mathbf{u}(t) \;\; = \;\; -\mathbf{K_P}\; \mathbf{e}(t) - \mathbf{K_I} \int \mathbf{e}(t)dt \qquad (5.2.6)$$

with the diagonal integral gain matrix $\mathbf{K_I}$. Defining

$$\dot{\boldsymbol{\epsilon}}(t) \;\; = \;\; \mathbf{e}(t), \qquad\qquad (5.2.7)$$

we arrive at

$$\boldsymbol{\tau}(t) \;\; = \;\; \mathbf{M}(\mathbf{q}(t)) \left[\ddot{\mathbf{q}}_{\mathbf{d}}(t) + \mathbf{K_P}\; \mathbf{e}(t) + \mathbf{K_I} \int \mathbf{e}(t)dt \right] + \mathbf{N}(\mathbf{q}(t), \dot{\mathbf{q}}(t)). \;\; (5.2.8)$$

5.2.3 PID Controller

Figure 5.3 shows the block diagram of a PID controller with control signal as

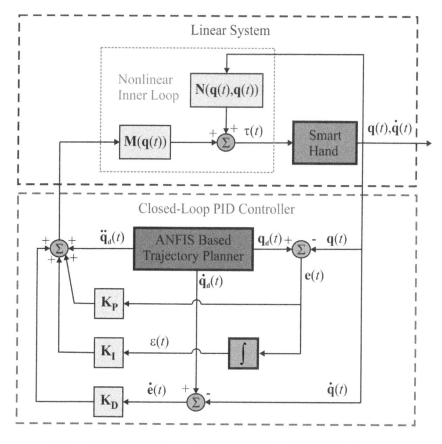

Figure 5.3   Block Diagram of Fusion PID Controller for Robotic Hand

$$\mathbf{u}(t) \quad = \quad -\mathbf{K_P}\ \mathbf{e}(t) - \mathbf{K_I} \int \mathbf{e}(t)dt - \mathbf{K_D}\ \dot{\mathbf{e}}(t). \tag{5.2.9}$$

We then rewrite (5.1.13) as

$$\boldsymbol{\tau}(t) \quad = \quad \mathbf{M}(\mathbf{q}(t)) \left[\ddot{\mathbf{q}}_d(t) + \mathbf{K_P}\ \mathbf{e}(t) + \mathbf{K_I} \int \mathbf{e}(t)dt + \mathbf{K_D}\ \dot{\mathbf{e}}(t) \right]$$
$$+ \mathbf{N}(\mathbf{q}(t), \dot{\mathbf{q}}(t)). \tag{5.2.10}$$

5.3   Optimal Controller

5.3.1   Optimal Regulation

Optimization is a very desirable feature in day-to-day life. The main objective of optimal control is to determine control signals that will cause a process (plant) to satisfy some physical constraints and at the same time extremize (maximize and minimize) a chosen performance criterion (performance index or cost function) [6].

The formulation of optimal control problem requires

1. a mathematical description (or model) of the process to be controlled (generally in state variable form),

2. a specification of the performance index, and

3. a statement of boundary conditions and the physical constraints on the states and/or controls.

5.3.2   Linear Quadratic Optimal Control with Tracking System

In obtaining the *linear* system (5.1.14) from the original nonlinear system (3.3.16), there has been no state-space transformation. Further, the difficult design of a controller for the original nonlinear system (3.3.16) has been transformed into a simple design of a controller for the linear system (5.1.15). If we select the control function $\mathbf{u}(t)$ to stabilize the linear system (5.1.14) and make the tracking error zero, then the nonlinear torque-control law $\boldsymbol{\tau}(t)$ given by (5.1.13) will command the robotic hand (3.3.16) to follow the desired trajectory $\mathbf{q}_d(t)$. With $\boldsymbol{\tau}(t)$ given by (5.1.13), the original robotic hand system (3.3.16) becomes

$$\mathbf{M}(\mathbf{q}(t))\ \ddot{\mathbf{q}}(t) + \mathbf{N}(\mathbf{q}(t), \dot{\mathbf{q}}(t)) = \mathbf{M}(\mathbf{q}(t))\left[\ddot{\mathbf{q}}_d(t) - \mathbf{u}(t)\right] + \mathbf{N}(\mathbf{q}(t), \dot{\mathbf{q}}(t)),$$
$$\text{that is, } \ddot{\mathbf{e}}(t) = \mathbf{u}(t), \tag{5.3.1}$$

which is exactly the linear system (5.1.14).

Our objective is to control the linear system (5.1.15) in such a way that the state variable $\mathbf{x}(t) = [\mathbf{q}'(t)\ \dot{\mathbf{q}}'(t)]'$ *tracks* the *desired* output $\mathbf{z}(t) = [\mathbf{q}_d'(t)\ \dot{\mathbf{q}}_d'(t)]'$ as close as possible during the interval $[t_0,\ t_f]$ with minimum control energy. For this, let us define the *error* vector as

$$\mathbf{e}(t) = \mathbf{z}(t) - \mathbf{x}(t), \tag{5.3.2}$$

and choose the performance index J [6] as

$$J = \frac{1}{2}\mathbf{e}'(t_f)\mathbf{F}(t_f)\mathbf{e}(t_f)$$
$$+ \frac{1}{2}\int_{t_0}^{t_f} \left[\mathbf{e}'(t)\mathbf{Q}\mathbf{e}(t) + \mathbf{u}'(t)\mathbf{R}\mathbf{u}(t)\right] dt. \tag{5.3.3}$$

We assume that $\mathbf{F}(t_f)$ and $\mathbf{Q}$ are symmetric, *positive semidefinite* matrices, and $\mathbf{R}$ is symmetric, *positive definite* matrix. We use Pontryagin minimum principle [6] and then solve the matrix differential Riccati equation (DRE)

$$\dot{\mathbf{P}}(t) = -\mathbf{P}(t)\mathbf{A} - \mathbf{A}'\mathbf{P}(t) + \mathbf{P}(t)\mathbf{BR}^{-1}\mathbf{B}'\mathbf{P}(t) - \mathbf{Q}, \qquad (5.3.4)$$

with final condition $\mathbf{P}(t_f) = \mathbf{F}(t_f)$, and the non-homogeneous vector differential equation

$$\dot{\mathbf{g}}(t) = -\left[\mathbf{A} - \mathbf{BR}^{-1}\mathbf{B}'\mathbf{P}(t)\right]' \mathbf{g}(t) - \mathbf{Qz}(t), \qquad (5.3.5)$$

with final condition $\mathbf{g}(t_f) = \mathbf{F}(t_f)\mathbf{z}(t_f)$. Then the optimal state $\mathbf{x}^*(t)$ is solved from

$$\dot{\mathbf{x}}^*(t) = \left[\mathbf{A} - \mathbf{BR}^{-1}\mathbf{B}'\mathbf{P}(t)\right]\mathbf{x}^*(t) + \mathbf{BR}^{-1}\mathbf{B}'\mathbf{g}(t) \qquad (5.3.6)$$

with initial condition $\mathbf{x}(t_0)$ and optimal control $\mathbf{u}^*(t)$ is calculated by

$$\mathbf{u}^*(t) = -\mathbf{R}^{-1}\mathbf{B}'\mathbf{P}(t)\mathbf{x}^*(t) + \mathbf{R}^{-1}\mathbf{B}'\mathbf{g}(t). \qquad (5.3.7)$$

Finally, the optimal required torque $\boldsymbol{\tau}^*(t)$ is obtained by

$$\boldsymbol{\tau}^*(t) = \mathbf{M}(\mathbf{q}(t))\mathbf{u}^*(t) + \mathbf{N}(\mathbf{q}(t), \dot{\mathbf{q}}(t)). \qquad (5.3.8)$$

Summarizing, Figure 5.4 shows the block diagram of a finite-time linear quadratic optimal controller tracking system for the robotic hand. Use of feedback linearization technique converts the nonlinear dynamics to linear. Then the closed-loop finite-time linear quadratic optimal controller through Pontryagin minimum principle is implemented to track the desired trajectory planning using cubic polynomial. $\mathbf{P}(t)$ and $\mathbf{g}(t)$ are computed by solving the matrix differential Riccati and the non-homogeneous vector differential equations with boundary conditions, respectively. Finally, the optimal state $\mathbf{x}^*(t)$ and optimal control $\mathbf{u}^*(t)$ are obtained in order to calculate the required torque $\boldsymbol{\tau}^*(t)$.

5.3.3   A Modified Optimal Control with Tracking System

Our previous works [5, 7] showed that the original optimal control can avoid overshooting and oscillation problems and get better results than GA-tuned PID control [5, 8], but this optimal control method takes execution time when applied to the robotic hand. To improve the performance of the original optimal controller, we change the performance index $\hat{J}$ [6] to include an exponential term as

$$\hat{J} = \frac{1}{2}e^{2\alpha t_f}\mathbf{e}'(t_f)\mathbf{F}(t_f)\mathbf{e}(t_f)$$

$$+ \frac{1}{2}\int_{t_0}^{t_f} e^{2\alpha t}\left[\mathbf{e}'(t)\mathbf{Qe}(t) + \mathbf{u}'(t)\mathbf{Ru}(t)\right]dt, \qquad (5.3.9)$$

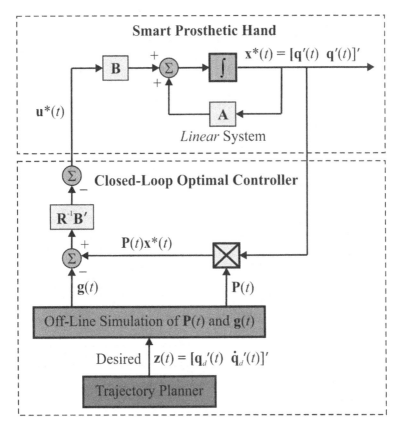

Figure 5.4   Block Diagram of Linear Quadratic Optimal Controller Tracking System for Robotic Hand

where α is a positive parameter. We need to find the optimal control which minimizes the new performance index $\hat{J}$ (5.3.9) under the dynamical constraint (5.1.15). This problem can be solved by modifying the original system, so the following transformations can be developed as

$$\hat{\mathbf{e}}(t) = e^{\alpha t}\mathbf{e}(t); \quad \hat{\mathbf{z}}(t) = e^{\alpha t}\mathbf{z}(t);$$
$$\hat{\mathbf{x}}(t) = e^{\alpha t}\mathbf{x}(t); \quad \hat{\mathbf{u}}(t) = e^{\alpha t}\mathbf{u}(t). \tag{5.3.10}$$

Then, using the transformations (5.3.10), it is easy to see that the *new* system becomes

$$\dot{\hat{\mathbf{x}}}(t) = \frac{d}{dt}\{e^{\alpha t}\mathbf{x}(t)\} = \alpha e^{\alpha t}\mathbf{x}(t) + e^{\alpha t}\dot{\mathbf{x}}(t)$$
$$= \alpha\hat{\mathbf{x}}(t) + e^{\alpha t}[\mathbf{A}\mathbf{x}(t) + \mathbf{B}\mathbf{u}(t)]$$
$$\dot{\hat{\mathbf{x}}}(t) = (\mathbf{A} + \alpha\mathbf{I})\hat{\mathbf{x}}(t) + \mathbf{B}\hat{\mathbf{u}}(t). \tag{5.3.11}$$

Considering the minimization of the modified system defined by (5.3.11) and (5.3.9), the new optimal control $\hat{u}^*(t)$, which is similar to (5.3.7), is given by

$$\hat{u}^*(t) = -R^{-1}B'\hat{P}(t)\hat{x}^*(t) + R^{-1}B'\hat{g}(t). \qquad (5.3.12)$$

Here, the matrix $\hat{P}(t)$ and the vector $\hat{g}(t)$ are respectively the solutions of DRE

$$
\begin{aligned}
\dot{\hat{P}}(t) &= -\hat{P}(t)(A + \alpha I) - (A' + \alpha I)\hat{P}(t) \\
&\quad +\hat{P}(t)BR^{-1}B'\hat{P}(t) - Q,
\end{aligned} \qquad (5.3.13)
$$

with final condition $\hat{P}(t_f) = F(t_f)$, and the non-homogeneous vector differential equation

$$\dot{\hat{g}}(t) = -\left[A + \alpha I - BR^{-1}B'\hat{P}(t)\right]' \hat{g}(t) - Q\hat{z}(t), \qquad (5.3.14)$$

with final condition $\hat{g}(t_f) = F(t_f)\hat{z}(t_f)$. Using the optimal control (5.3.12) in the new system (5.3.11), we get the optimal closed-loop system as

$$\dot{\hat{x}}^*(t) = \left[A + \alpha I - BR^{-1}B'\hat{P}(t)\right] \hat{x}^*(t) + BR^{-1}B'\hat{g}(t) \qquad (5.3.15)$$

with initial condition $\hat{x}(t_0)$.

Hence, applying the transformations (5.3.10) in the new system (5.3.15), the optimal control of the original system (5.1.15) and the associated performance measure (5.3.9) is given by

$$
\begin{aligned}
u^*(t) &= e^{-\alpha t}\hat{u}^*(t) = -e^{-\alpha t}R^{-1}B' \left[\hat{P}(t)\hat{x}^*(t) - \hat{g}(t)\right] \\
&= -R^{-1}B'\hat{P}(t)x^*(t) + e^{-\alpha t}R^{-1}B'\hat{g}(t).
\end{aligned} \qquad (5.3.16)
$$

Interestingly, this desired (original) optimal control has the same matrix DRE solutions $\hat{P}(t) = P(t)$ as the optimal control of the new system with $\hat{g}(t) = e^{\alpha t}g(t)$ compared with (5.3.16) and (5.3.7). We see that the closed-loop optimal control system (5.3.15) has eigenvalues with real parts less than $-\alpha$. In other words, the state $x^*(t)$ approaches zero at least as fast as $e^{-\alpha t}$ [9].

5.4 Adaptive Controller

Adaptive control involves modifying the control law used by a controller to cope with the fact that the parameters of the system being controlled are slowly time varying or uncertain. For example, as an aircraft flies, its mass will slowly decrease as a result of fuel consumption; a control law that adapts itself to such changing conditions is needed. Adaptive control is different from robust control in the sense that it does not need a priori information about the bounds on these uncertain or time-varying parameters; robust control guarantees that if the changes are within given bounds the control law need not be changed, while adaptive control is precisely concerned with control law changes.

Applying adaptive control to the robotic hand, the response with adaptive controller works well even if the masses of robots are unknown by the controller. After the initial errors, the actual joint angles closely match the desired joint angles. In adaptive control, the controller dynamics allows for learning of the unknown parameters, so that the performance improves over time [2].

The tracking error $\mathbf{e}(t)$ and the filtered tracking error $\mathbf{r}(t)$ are defined as

$$\mathbf{e}(t) = \mathbf{q}_d(t) - \mathbf{q}(t), \tag{5.4.1}$$

$$\mathbf{r}(t) = \dot{\mathbf{e}}(t) + \mathbf{\Lambda}\mathbf{e}(t). \tag{5.4.2}$$

Here, $\mathbf{q}_d(t)$ is the *desired* angle vector of joints; $\mathbf{q}(t)$ is the *actual* angle vector of joints; $\mathbf{\Lambda} = diag(\lambda_1, \lambda_2, \ldots, \lambda_n)$ is the positive definite diagonal gain matrix. The filtered error (5.4.2) ensures stability of the overall system so that the tracking error (5.4.1) is bounded. Figure 5.5 shows the block diagram of the adaptive con-

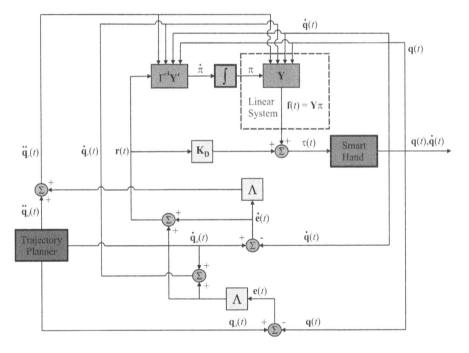

Figure 5.5 Block Diagram of Adaptive Controller for Robotic Hand

troller: tracking errors $\mathbf{e}(t)$ are calculated by actual angles $\mathbf{q}(t)$ and desired angles $\mathbf{q}_d(t)$, which are based on trajectory planner. Then filtered tracking errors $\mathbf{r}(t)$ are computed by error changes and the parameters $\mathbf{\Lambda}$ multiplying errors. The required torque $\boldsymbol{\tau}(t)$ of the robotic hand nonlinear system is computed by the nonlinear term $\mathbf{f}(t)$ and the gain $\mathbf{K_D}$ multiplying the filtered tracking errors.

Differentiating and substituting (5.4.2) into (5.1.1) gives the dynamic equation in terms of the filtered error $\mathbf{r}(t)$ as

$$\mathbf{M}(\mathbf{q}(t))\dot{\mathbf{r}}(t) = -\mathbf{C}_m(\mathbf{q}(t), \dot{\mathbf{q}}(t))\mathbf{r}(t) + \mathbf{f}(t) - \boldsymbol{\tau}(t), \tag{5.4.3}$$

where $\mathbf{C}(\mathbf{q}(t), \dot{\mathbf{q}}(t)) = \mathbf{C}_m(\mathbf{q}(t), \dot{\mathbf{q}}(t))\dot{\mathbf{q}}(t)$ and the nonlinear term $\mathbf{f}(t)$ is defined as

$$
\begin{aligned}
\mathbf{f}(t) &= \mathbf{M}(\mathbf{q}(t))(\ddot{\mathbf{q}}_d(t) + \boldsymbol{\Lambda}\dot{\mathbf{e}}(t)) + \mathbf{G}(\mathbf{q}(t)) + \\
&\quad \mathbf{C}_m(\mathbf{q}(t), \dot{\mathbf{q}}(t))(\dot{\mathbf{q}}_d(t) + \boldsymbol{\Lambda}\mathbf{e}(t)) + \boldsymbol{\tau}_{dis}, \\
&= \mathbf{Y}\boldsymbol{\pi}.
\end{aligned} \tag{5.4.4}
$$

Here, $\boldsymbol{\tau}_{dis}$ is the unknown disturbance. $\mathbf{Y}$ is a regression matrix of known robot functions and $\boldsymbol{\pi}$ is a vector of unknown parameters [10]. The regression matrix $\mathbf{Y}$ and the unknown parameter vector $\boldsymbol{\pi}$ of two-link thumb and three-link index finger are given in Appendix 5.A [11]. The torque vector $\boldsymbol{\tau}(t)$ is calculated by

$$\boldsymbol{\tau}(t) = \mathbf{f}(t) + \mathbf{K_D}\mathbf{r}(t). \tag{5.4.5}$$

The unknown parameter rate vector $\dot{\boldsymbol{\pi}}$ is updated by

$$\dot{\boldsymbol{\pi}} = \boldsymbol{\Gamma}^{-1}\mathbf{Y}'\mathbf{r}(t) \tag{5.4.6}$$

where $\boldsymbol{\Gamma}$ is a tuning parameter diagonal matrix.

5.5 Simulation Results and Discussion

This section presents simulations with the PID, optimal and adaptive controllers for the two-link thumb and three-link index finger of a smart robotic hand. Then, the simulations with the PID, optimal and adaptive controllers for 14-DOF, five-fingered robotic hand will be presented.

5.5.1 Two Link Thumb

The various parameters [12] relating to desired trajectory and the two-link thumb selected for the simulations are given in Table 5.1 and the side length of the target square-shaped object is 0.010 (m) as shown in Figure 2.16. PID diagonal coefficients, $\mathbf{K_P}(t)$, $\mathbf{K_I}(t)$, and $\mathbf{K_D}(t)$, are 100. As for optimal control coefficients, $\mathbf{A}$, $\mathbf{B}$, $\mathbf{R}(t)$, and $\mathbf{Q}^t(t)$ of thumb are chosen as

$$
\begin{aligned}
\mathbf{A} &= \begin{bmatrix} 0 & \mathbf{I} \\ 0 & 0 \end{bmatrix}, \quad \mathbf{B} = \begin{bmatrix} 0 \\ \mathbf{I} \end{bmatrix}, \quad \mathbf{R}(t) = \frac{1}{30}\mathbf{I}, \quad \mathbf{Q}^t(t) = \begin{bmatrix} \mathbf{Q}_{11} & \mathbf{Q}_{12} \\ \mathbf{Q}_{12} & \mathbf{Q}_{22} \end{bmatrix}, \\
\mathbf{Q}_{11} &= \begin{bmatrix} 10 & 2 \\ 2 & 10 \end{bmatrix}, \quad \mathbf{Q}_{22} = \begin{bmatrix} 30 & 0 \\ 0 & 30 \end{bmatrix}, \quad \mathbf{Q}_{12} = \begin{bmatrix} -4 & 4 \\ 3 & -6 \end{bmatrix}.
\end{aligned} \tag{5.5.1}
$$

Table 5.1   Parameter Selection of Thumb

Parameters	Values
Thumb	
Time (t_0, t_f)	$0, 20$ (sec)
Desired initial position (X_0^t, Y_0^t)	$0.035, 0.060$ (m)
Desired final position (X_f^t, Y_f^t)	$0.0495, 0.060$ (m)
Desired initial velocity $(\dot{X}_0^t, \dot{Y}_0^t)$	$0, 0$ (m/s)
Desired final velocity $(\dot{X}_f^t, \dot{Y}_f^t)$	$0, 0$ (m/s)
Length (L_1^t, L_2^t)	$0.040, 0.040$ (m)
Mass (m_1^t, m_2^t)	$0.043, 0.031$ (kg)
Inertia (I_{zz1}^t, I_{zz2}^t)	$6.002 \times 10^{-6}, 4.327 \times 10^{-6}$ (kg-m^2)

Figure 5.6 shows the simulation with PID controller while Figure 5.7 shows the simulation with the presented finite-time optimal control method. It is clearly seen that the results using proposed optimal control method overcome the overshoot and oscillation problems. Next, we present simulations with a PID controller and adaptive

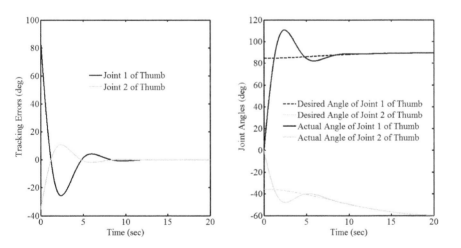

Figure 5.6   Tracking Errors and Joint Angles of PID Controller for Thumb

controller for the two-link thumb of a smart robotic hand. Figure 5.8 shows the PID control and adaptive control methods for two-link thumb [13].

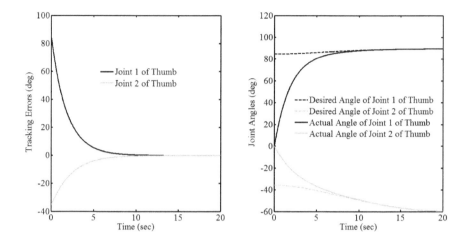

Figure 5.7 Tracking Errors and Joint Angles of Optimal Control for Thumb

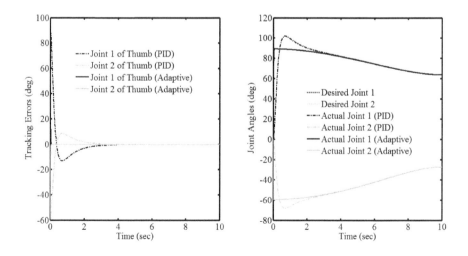

Figure 5.8 Tracking Errors and Joint Angles of PID and Adaptive Controllers for Thumb

5.5.2 Three Link Index Finger

Similarly, the various parameters [12] relating to desired trajectory and the three-link index finger selected for the simulations are given in Table 5.2 and the side length of

Table 5.2 Parameter Selection of Index Finger

Parameters	Values
Index Finger	
Time (t_0, t_f)	0, 20 (sec)
Desired initial position (X_0^i, Y_0^i)	0.065, 0.080 (m)
Desired final position (X_f^i, Y_f^i)	0.010, 0.060 (m)
Desired initial velocity $(\dot{X}_0^i, \dot{Y}_0^i)$	0, 0 (m/s)
Desired final velocity $(\dot{X}_f^i, \dot{Y}_f^i)$	0, 0 (m/s)
Length (L_1^i, L_2^i, L_3^i)	0.040, 0.040, 0.030
Mass (m_1^i, m_2^i, m_3^i)	0.045, 0.025, 0.017 (kg)
Inertia $(I_{zz1}^i, I_{zz2}^i, I_{zz3}^i)$	$9.375 \times 10^{-6}, 3.333 \times 10^{-6}, 1.125 \times 10^{-6}$ (kg-m^2)
distance (d)	0.035 (m)

the target square-shaped object is 0.010 (m) as shown in Figure 2.16. PID diagonal coefficients, $\mathbf{K_P}(t)$, $\mathbf{K_I}(t)$, and $\mathbf{K_D}(t)$, are 100 and optimal control coefficients, $\mathbf{A}$, $\mathbf{B}$, $\mathbf{R}(t)$, and $\mathbf{Q}^i(t)$ of index finger are chosen as

$$\mathbf{A} = \begin{bmatrix} 0 & \mathbf{I} \\ 0 & 0 \end{bmatrix}, \quad \mathbf{B} = \begin{bmatrix} 0 \\ \mathbf{I} \end{bmatrix}, \quad \mathbf{R}(t) = \frac{1}{30}\mathbf{I}, \quad \mathbf{Q}^i(t) = \begin{bmatrix} \mathbf{Q_{11}} & \mathbf{Q_{12}} & \mathbf{Q_{13}} \\ \mathbf{Q_{12}} & \mathbf{Q_{22}} & \mathbf{Q_{23}} \\ \mathbf{Q_{13}} & \mathbf{Q_{23}} & \mathbf{Q_{33}} \end{bmatrix},$$

$$\mathbf{Q_{11}} = \begin{bmatrix} 10 & 2 \\ 2 & 10 \end{bmatrix}, \quad \mathbf{Q_{22}} = \begin{bmatrix} 30 & 0 \\ 0 & 30 \end{bmatrix}, \quad \mathbf{Q_{33}} = \begin{bmatrix} 20 & 1 \\ 1 & 20 \end{bmatrix},$$

$$\mathbf{Q_{12}} = \begin{bmatrix} -4 & 4 \\ 3 & -6 \end{bmatrix}, \quad \mathbf{Q_{13}} = \begin{bmatrix} -4 & 4 \\ 3 & -6 \end{bmatrix}, \quad \mathbf{Q_{23}} = \begin{bmatrix} -4 & 3 \\ 4 & -6 \end{bmatrix}.$$

Figure 5.9 shows the simulation with PID controller while Figure 5.10 shows the simulations with the presented finite-time optimal control method. Similar to the results of the thumb, it is clearly seen that the results using proposed optimal control method can overcome the overshoot and oscillation problems.

The desired path 2 for three-link index finger is

$$\mathbf{q}_d^i = [A_1^i \sin(\frac{2\pi}{T}) \quad A_2^i \sin(\frac{2\pi}{T}) \quad 0.7A_2^i \sin(\frac{2\pi}{T})]'. \tag{5.5.2}$$

The various parameters [12] relating to desired trajectory and the two-link thumb/three-link index finger selected for the simulations are given as: $A_1^i = A_2^i$

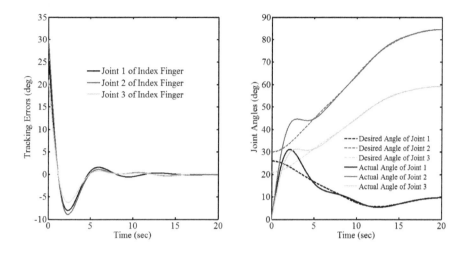

Figure 5.9   Tracking Errors and Joint Angles of PID Controller for Index Finger

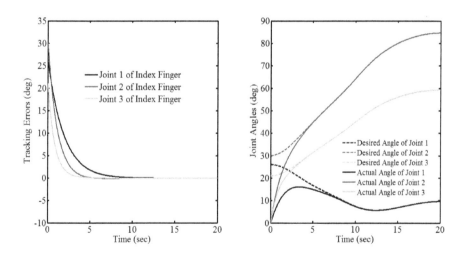

Figure 5.10   Tracking Errors and Joint Angles of Optimal Control for Index Finger

= 0.1; T = 2 (path 2); initial position (X^t, Y^t) = (0.035, 0.060) and final position (X^t, Y^t) = (0.0495, 0.060) (m); initial and final velocities are zero (path 1); the lengths of the links 1, 2, and 3 are 0.040, 0.040, and 0.030 (m). PID diagonal coefficients, $\mathbf{K_P}(t)$, $\mathbf{K_I}(t)$ and $\mathbf{K_D}(t)$, are 100. As for the adaptive diagonal coefficients, $\mathbf{K_D}$, $\mathbf{\Lambda}$, and $\mathbf{\Gamma}$ are also chosen as 100. Figures 5.8 and 5.11 show our previous work [14, 15] with PID controller and the adaptive control method for two-link thumb and three-link index finger. It is clearly seen that the results using proposed fusion adap-

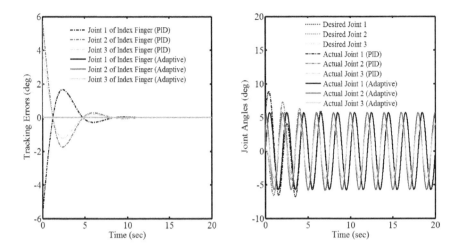

Figure 5.11   Tracking Errors and Joint Angles of PID and Adaptive Controllers for Path 2 of Index Finger

tive control strategy overcomes the overshooting and oscillation problems even if the robotic hand is under disturbances [16].

5.5.3   Three Dimensional Five Fingered Robotic Hand

Next, we present simulations with a PID controller and a finite-time linear quadratic optimal controller for the 14-DOF, five-fingered smart robotic hand.

5.5.3.1   PID Control   The parameters of the two-link thumb/three-link index finger [12] were related to desired trajectory. All parameters of the smart robotic hand selected for the simulations are given in Table 5.3 and the side length and length of the target rectangular rod are 0.010 and 0.100 (m), respectively, as shown in Figure 2.17. The relating parameters between the global coordinate and the local coordinates are defined in Table 5.4 [11, 17]. In addition, all links are assumed as circular cylinder with the radius (R) 0.010 (m), so the inertia I_{zzk}^{j} of each link k of each finger j ($j = t, i, m, r$, and l) is calculated as

$$I_{zzk}^{j} \;=\; \frac{1}{4}m_{k}^{j}R^{2} + \frac{1}{3}m_{k}^{j}L_{k}^{j}{}^{2}. \qquad (5.5.3)$$

All initial actual angles are zero and all PID diagonal coefficients, $\mathbf{K_P}(t)$, $\mathbf{K_I}(t)$, and $\mathbf{K_D}(t)$, are 100.

5.5.3.2 *Optimal Control* Optimal control coefficients, $\mathbf{A}$, $\mathbf{B}$, $\mathbf{F}(t_f)$, $\mathbf{R}(t)$, and $\mathbf{Q}(t)$ of all fingers are chosen as

$$
\mathbf{A} = \begin{bmatrix} 0 & \mathbf{I} \\ 0 & 0 \end{bmatrix}, \quad \mathbf{B} = \begin{bmatrix} 0 \\ \mathbf{I} \end{bmatrix}, \quad \mathbf{F}(t_f) = 0, \quad \mathbf{R}(t) = \frac{1}{30}\mathbf{I},
$$

$$
\mathbf{Q}^t(t) = \begin{bmatrix} \mathbf{Q}_{11} & \mathbf{Q}_{12} \\ \mathbf{Q}_{12} & \mathbf{Q}_{22} \end{bmatrix}, \quad \mathbf{Q}^j(t) = \begin{bmatrix} \mathbf{Q}_{11} & \mathbf{Q}_{12} & \mathbf{Q}_{13} \\ \mathbf{Q}_{12} & \mathbf{Q}_{22} & \mathbf{Q}_{23} \\ \mathbf{Q}_{13} & \mathbf{Q}_{23} & \mathbf{Q}_{33} \end{bmatrix},
$$

$$
\mathbf{Q}_{11} = \begin{bmatrix} 10 & 2 \\ 2 & 10 \end{bmatrix}, \quad \mathbf{Q}_{22} = \begin{bmatrix} 30 & 0 \\ 0 & 30 \end{bmatrix}, \quad \mathbf{Q}_{33} = \begin{bmatrix} 20 & 1 \\ 1 & 20 \end{bmatrix},
$$

$$
\mathbf{Q}_{12} = \begin{bmatrix} -4 & 4 \\ 3 & -6 \end{bmatrix}, \quad \mathbf{Q}_{13} = \begin{bmatrix} -4 & 4 \\ 3 & -6 \end{bmatrix}, \quad \mathbf{Q}_{23} = \begin{bmatrix} -4 & 3 \\ 4 & -6 \end{bmatrix}.
$$

Here $j = i$, m, r, and l. The first term of the right side of (5.3.2) can be neglected by using $\mathbf{F}(t_f)$ as the zero matrix. In this example, there is no significant difference.

Figures 5.12 – 5.16 show the simulations with fusion PID controller and fusion optimal controller. It is clearly seen in the superiority of the fusion optimal controller in suppressing the overshoots and transients associated with the fusion PID controller.

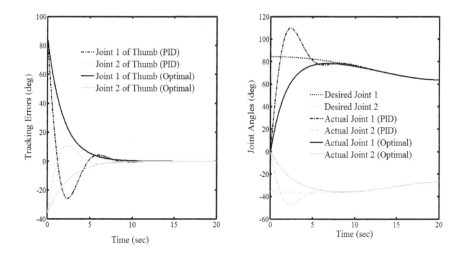

Figure 5.12 Tracking Errors and Joint Angles of PID and Optimal Controllers for Thumb of Five-Fingered Robotic Hand

Table 5.3   Parameter Selection of the Smart Robotic Hand

Parameters	Values
Thumb	
Time $(t_0, t_f)^*$	0, 20 (sec)
Desired initial position $(X_0^t, Y_0^t)^{**}$	0.035, 0.060 (m)
Desired final position $(X_f^t, Y_f^t)^{**}$	0.0495, 0.060 (m)
Desired initial velocity $(\dot{X}_0^t, \dot{Y}_0^t)^*$	0, 0 (m/s)
Desired final velocity $(\dot{X}_f^t, \dot{Y}_f^t)^*$	0, 0 (m/s)
Length (L_1^t, L_2^t)	0.040, 0.040 (m)
Mass (m_1^t, m_2^t)	0.043, 0.031 (kg)
Index Finger	
Desired initial position $(X_0^i, Y_0^i)^{**}$	0.065, 0.080 (m)
Desired final position $(X_f^i, Y_f^i)^{**}$	0.010, 0.060 (m)
Length (L_1^i, L_2^i, L_3^i)	0.040, 0.040, 0.030 (m)
Mass (m_1^i, m_2^i, m_3^i)	0.045, 0.025, 0.017 (kg)
Middle Finger	
Desired initial position $(X_0^m, Y_0^m)^{**}$	0.065, 0.080 (m)
Desired final position $(X_f^m, Y_f^m)^{**}$	0.005, 0.060 (m)
Length (L_1^m, L_2^m, L_3^m)	0.044, 0.044, 0.033 (m)
Mass (m_1^m, m_2^m, m_3^m)	0.050, 0.028, 0.017 (kg)
Ring Finger	
Desired initial position $(X_0^r, Y_0^r)^{**}$	0.065, 0.080 (m)
Desired final position $(X_f^r, Y_f^r)^{**}$	0.010, 0.060 (m)
Length (L_1^r, L_2^r, L_3^r)	0.040, 0.040, 0.030 (m)
Mass (m_1^r, m_2^r, m_3^r)	0.041, 0.023, 0.014 (kg)
Little Finger	
Desired initial position $(X_0^l, Y_0^l)^{**}$	0.055, 0.080 (m)
Desired final position $(X_f^l, Y_f^l)^{**}$	0.020, 0.060 (m)
Length (L_1^l, L_2^l, L_3^l)	0.036, 0.036, 0.027 (m)
Mass (m_1^l, m_2^l, m_3^l)	0.041, 0.023, 0.014 (kg)

* All fingers use the same parameters.
** All parameters are in local coordinates.

Table 5.4   Parameter Selection of the Relation between Global and Local Frames

Parameters	Values
α	90 (deg)
β	45 (deg)
$\mathbf{d}^i$	$(0.035, 0, 0)$ (m)
$\mathbf{d}^m$	$(0.040, 0, -0.020)$ (m)
$\mathbf{d}^r$	$(0.035, 0, -0.040)$ (m)
$\mathbf{d}^l$	$(0.025, 0, -0.060)$ (m)

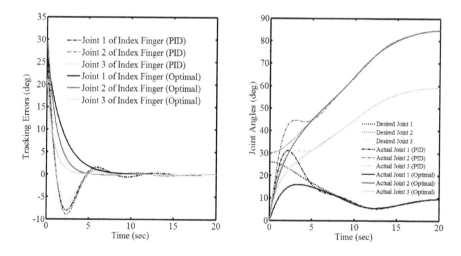

Figure 5.13   Tracking Errors and Joint Angles of PID and Optimal Controllers for Index Finger of Five-Fingered Robotic Hand

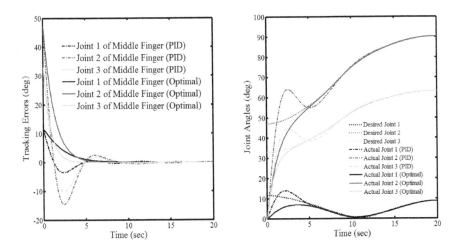

Figure 5.14   Tracking Errors and Joint Angles of PID and Optimal Controllers for Middle Finger of Five-Fingered Robotic Hand

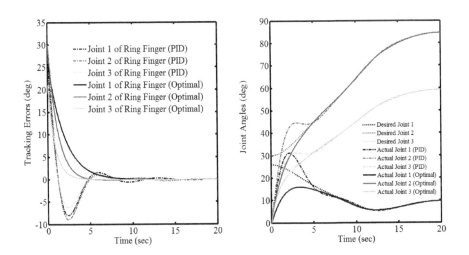

Figure 5.15   Tracking Errors and Joint Angles of PID and Optimal Controllers for Ring Finger of Five-Fingered Robotic Hand

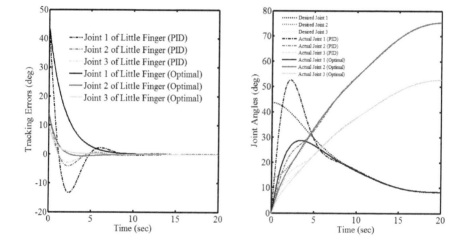

Figure 5.16 Tracking Errors and Joint Angles of PID and Optimal Controllers for Little Finger of Five-Fingered Robotic Hand

Figures 5.17, 5.19, 5.21, 5.23, and 5.25 show the tracking errors of thumb, index, middle, ring, and little fingers for the proposed five-fingered smart robotic hand, respectively. Figures 5.18, 5.20, 5.22, 5.24, and 5.26 show the desired/actual angles of thumb, index, middle, ring, and little fingers for the proposed five-fingered smart robotic hand, respectively. The observation that all tracking errors dramatically drop within 1 second and are less than 1 degree after convergence provides the evidence that the adaptive controller for the 14-DOF robotic hand enhances performance. The other observation that after convergence, all three-link fingers show more unstable errors than two-link thumb suggests that the more DOFs increase the difficulty of the adaptive controller without knowing the mass and inertia of the links of all fingers.

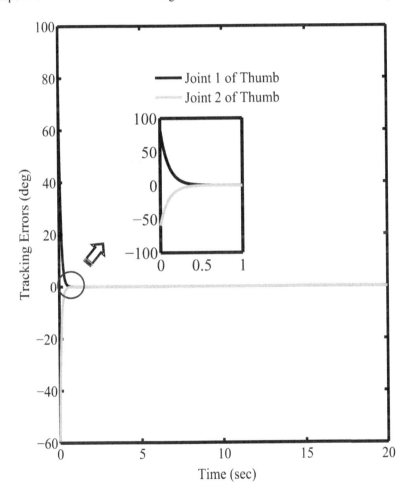

Figure 5.17   Tracking Errors of Adaptive Controller for Two-Link Thumb

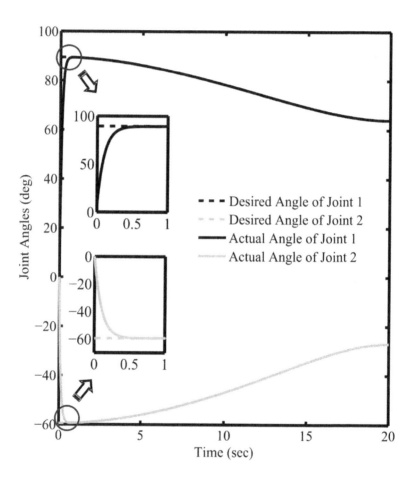

Figure 5.18 Tracking Angles of Adaptive Controller for Two-Link Thumb

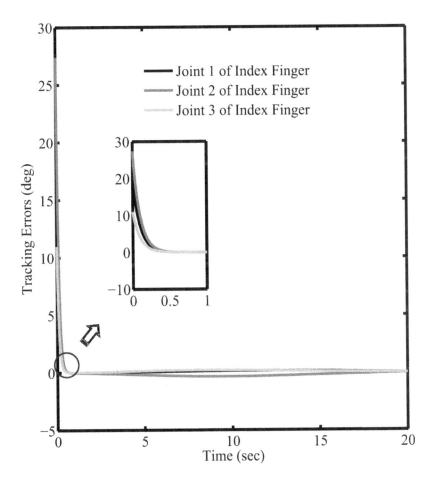

Figure 5.19 Tracking Errors of Adaptive Controller for Three-Link Index Finger

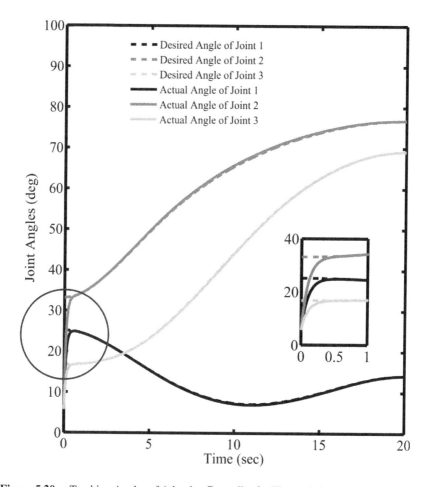

Figure 5.20 Tracking Angles of Adaptive Controller for Three-Link Index Finger

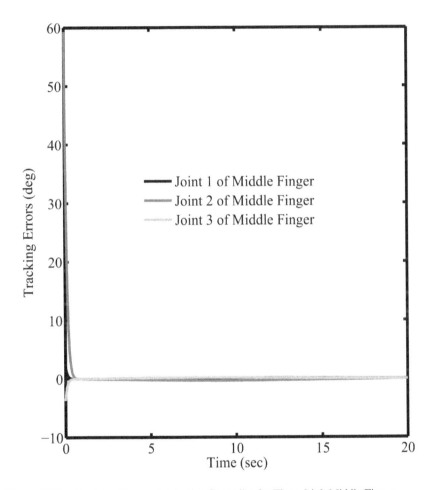

Figure 5.21   Tracking Errors of Adaptive Controller for Three-Link Middle Finger

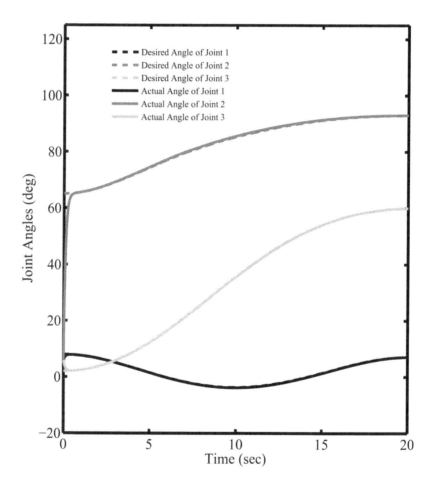

Figure 5.22   Tracking Angles of Adaptive Controller for Three-Link Middle Finger

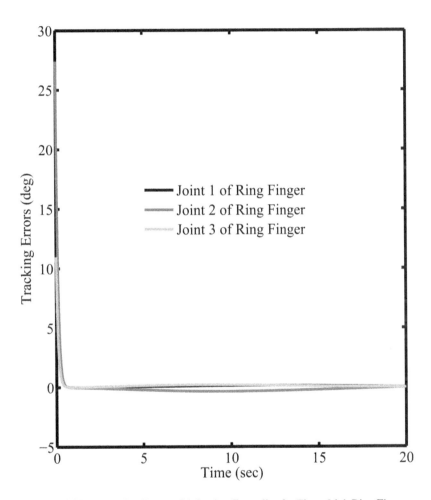

Figure 5.23   Tracking Errors of Adaptive Controller for Three-Link Ring Finger

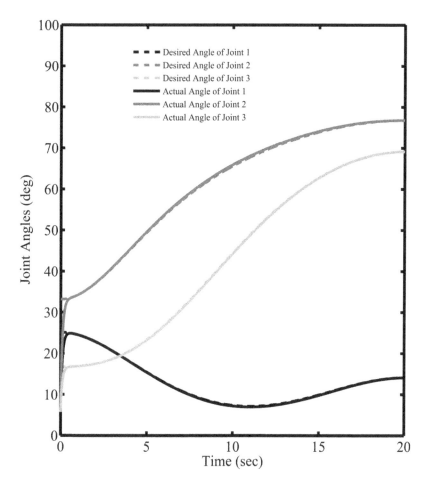

Figure 5.24   Tracking Angles of Adaptive Controller for Three-Link Ring Finger

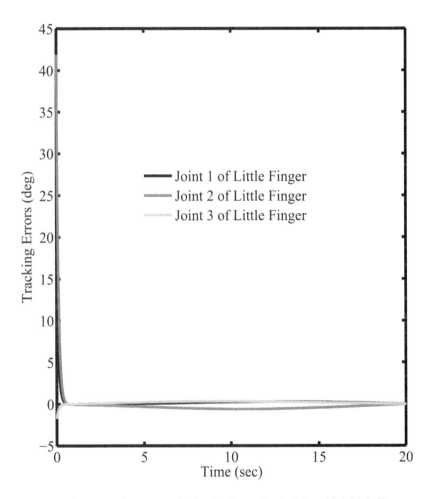

Figure 5.25   Tracking Errors of Adaptive Controller for Three-Link Little Finger

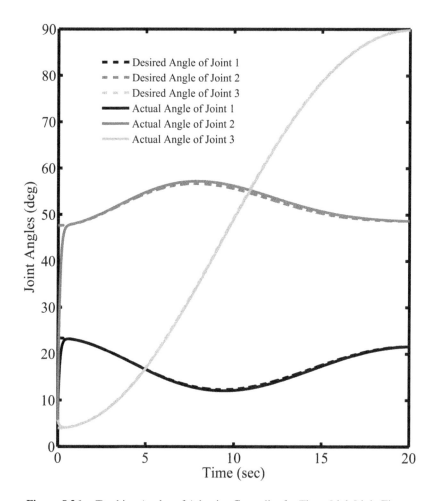

Figure 5.26   Tracking Angles of Adaptive Controller for Three-Link Little Finger

To compare the performance of the GA-tuned PID (see Section 6.2) and modified optimal controllers, Figures 5.27 and 5.28 show desired/actual angles and tracking errors of joints 1 and 2 for two-link thumb, respectively. GA-tuned PID control shows an overshooting problem. The problem is overcome by the original optimal control ($\alpha = 0$), but it takes at least 10 seconds for both joints. The performance is improved by the proposed optimal controller as the parameter α increases from 1 to 10. In other words, the convergence time is reduced to approximate 0.2 second as α is 10. For three-link index finger, the GA-tuned PID control causes not only overshooting but also oscillation problems as shown in Figures 5.29(a) and 5.30(a).

The optimal control with modified performance index ($\hat{J}$) embedded with an exponential term (α) also overcomes the overshooting and oscillation problems and obtains faster convergence speed as α increases. Similar simulations are also made for other three-link fingers. Taken together, these data suggest that the modified optimal control has higher accuracy and faster convergence speed than GA-tuned PID control [9].

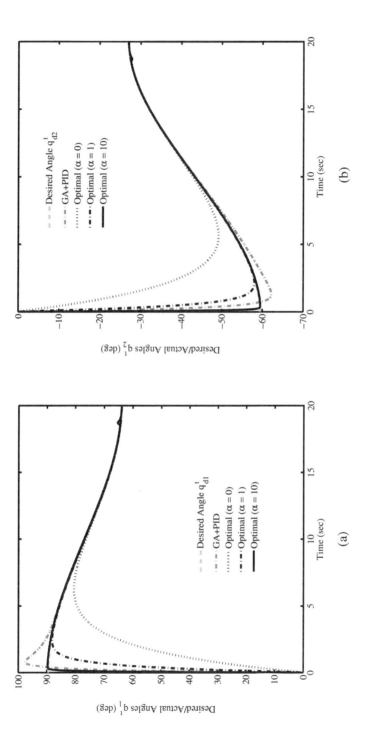

Figure 5.27   Desired/Actual Angular Positions of Two-Link Thumb: The Actual Angles (a) q_1^t and (b) q_2^t Regulated by GA-Tuned PID Controller and Modified Optimal Controller with the Different Parameters α ($\alpha = 0$, 1, and 10) Are Designed to Track the Desired Angles q_{d1}^t and q_{d2}^t.

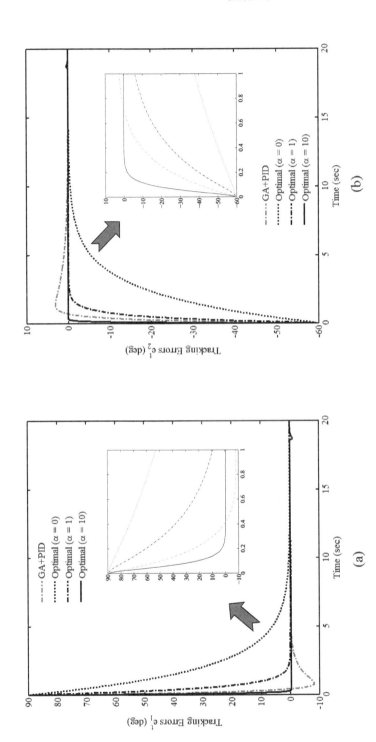

Figure 5.28 Tracking Errors of Two-Link Thumb: The Tracking Errors (a) e_1^t and (b) e_2^t Show that PID Controller with GA-Tuned Parameters Has an Overshooting Problem, Which Is Overcome by the Proposed Modified Optimal Controller with $\alpha = 0, 1,$ and 10.

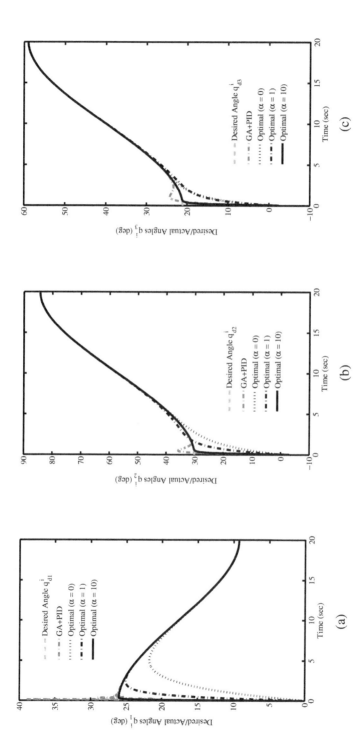

Figure 5.29 Desired/Actual Angular Positions of Three-Link Index Finger: The Actual Angles (a) q_1^i, (b) q_2^i, and (c) q_3^i Regulated by GA-Tuned PID Controller and Proposed Optimal Controller with the Different Parameters α ($\alpha = 0$, 1, and 10) Are Designed to Track the Desired Angles q_{d1}^i, q_{d2}^i, and q_{d3}^i, Respectively.

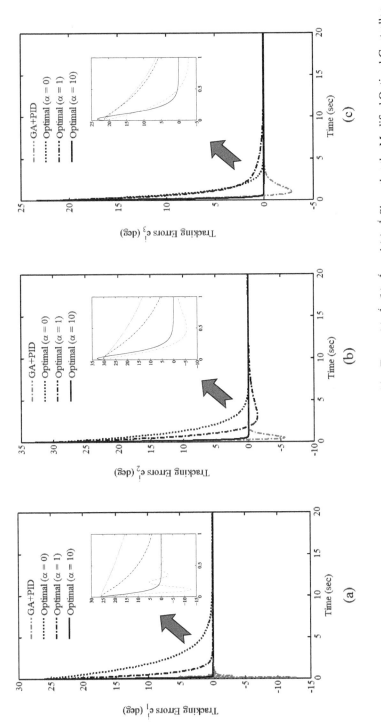

Figure 5.30   Tracking Errors of Three-Link Index Finger: The Tracking Errors (a) e_1^i, (b) e_2^i, and (c) e_3^i Show that the Modified Optimal Controller Acts Faster than GA-Based PID Controller as the Parameter α Increases from 0 to 10.

5.A Appendix: Regression Matrix

In Section 5.4, the regression matrix $\mathbf{Y}^t$ and the unknown parameter vector $\boldsymbol{\pi}^t$ of two-link thumb is expressed as

$$\mathbf{Y}^t = \begin{bmatrix} Y_{11}^t & Y_{12}^t & Y_{13}^t & Y_{14}^t \\ Y_{21}^t & Y_{22}^t & Y_{23}^t & Y_{24}^t \end{bmatrix},$$

$$\boldsymbol{\pi}^t = \begin{bmatrix} m_1^t & m_2^t & I_{zz1}^t & I_{zz2}^t \end{bmatrix}',$$

where

$$
\begin{aligned}
Y_{11}^t &= l_1^t l_1^t (\ddot{q}_{d1} + \lambda_1 \dot{e}_1) + g l_1^t C_1, \\
Y_{12}^t &= (2L_1^t l_2^t C_2 + L_1^t L_1^t + l_2^t l_2^t)(\ddot{q}_{d1} + \lambda_1 \dot{e}_1) + (L_1^t l_2^t C_2 + l_2^t l_2^t)(\ddot{q}_{d2} + \lambda_2 \dot{e}_2) \\
&\quad - L_1^t l_2^t S_2 \dot{q}_2 (\dot{q}_{d1} + \lambda_1 e_1) - L_1^t l_2^t S_2 (\dot{q}_1 + \dot{q}_2)(\dot{q}_{d2} + \lambda_2 e_2) + g l_2^t C_{12}, \\
Y_{13}^t &= \ddot{q}_{d1} + \lambda_1 \dot{e}_1, \\
Y_{14}^t &= \ddot{q}_{d1} + \lambda_1 \dot{e}_1 + \ddot{q}_{d2} + \lambda_2 \dot{e}_2, \\
Y_{21}^t &= 0, \\
Y_{22}^t &= (L_1^t l_2^t C_2 + l_2^t l_2^t)(\ddot{q}_{d1} + \lambda_1 \dot{e}_1) + l_2^t l_2^t (\ddot{q}_{d2} + \lambda_2 \dot{e}_2) \\
&\quad + L_1^t l_2^t S_2) \dot{q}_1 (\dot{q}_{d1} + \lambda_1 e_1) - (L_1^t l_2^t S_2) \dot{q}_1 (\dot{q}_{d2} + \lambda_2 e_2) + g l_2^t C_{12}, \\
Y_{23}^t &= 0, \\
Y_{24}^t &= \ddot{q}_{d1} + \lambda_1 \dot{e}_1 + \ddot{q}_{d2} + \lambda_2 \dot{e}_2, \\
C_1 &= \cos(q_1), \\
C_2 &= \cos(q_2), \\
S_2 &= \sin(q_2), \\
C_{12} &= \cos(q_1 + q_2).
\end{aligned}
$$

Similarly, the regression matrix $\mathbf{Y}^i$ and the unknown parameter vector $\boldsymbol{\pi}^i$ of three-link index finger are written as

$$\mathbf{Y}^i = \begin{bmatrix} Y_{11}^i & Y_{12}^i & Y_{13}^i & Y_{14}^i & Y_{15}^i & Y_{16}^i \\ Y_{21}^i & Y_{22}^i & Y_{23}^i & Y_{24}^i & Y_{25}^i & Y_{26}^i \\ Y_{31}^i & Y_{32}^i & Y_{33}^i & Y_{34}^i & Y_{35}^i & Y_{36}^i \end{bmatrix},$$

$$\boldsymbol{\pi}^i = \begin{bmatrix} m_1^i & m_2^i & m_3^i & I_{zz1}^i & I_{zz2}^i & I_{zz3}^i \end{bmatrix}'.$$

Here,

$$Y_{11}^i = l_1^i l_1^i (\ddot{q}_{d1} + \lambda_1 \dot{e}_1) + g l_1^i C_1 + g l_2^i C_{12},$$

$$Y_{12}^i = (2L_1^i l_2^i S_1 S_{12} + 2L_1^i l_2^i C_1 C_{12} + L_1^i L_1^i + l_2^i l_2^i)(\ddot{q}_{d1} + \lambda_1 \dot{e}_1)$$
$$+ (L_1^i l_2^i S_1 S_{12} + L_1^i l_2^i C_1 C_{12} + l_2^i l_2^i)(\ddot{q}_{d2} + \lambda_2 \dot{e}_2) + g L_1 C_1$$
$$+ (L_1^i l_2^i S_1 C_{12} - L_1^i l_2^i C_1 S_{12})\dot{q}_1(\dot{q}_{d2} + \lambda_2 e_2)$$
$$+ (L_1^i l_2^i S_1 C_{12} - L_1^i l_2^i C_1 S_{12})\dot{q}_2(\dot{q}_{d1} + \lambda_1 e_1)$$
$$+ (L_1^i l_2^i S_1 C_{12} - L_1^i l_2^i C_1 S_{12})\dot{q}_2(\dot{q}_{d2} + \lambda_2 e_2),$$

$$Y_{13}^i = (2L_1^i L_2^i S_1 S_{12} + 2L_1^i L_2^i C_1 C_{12} + 2L_1^i l_3^i S_1 S_{123} + 2L_1^i l_3^i C_1 C_{123}$$
$$+ 2L_2^i l_3^i S_{12} S_{123} + 2L_2^i l_3^i C_{12} C_{123} + L_1^i L_1^i + L_2^i L_2^i + l_3^i l_3^i)(\ddot{q}_{d1} + \lambda_1 \dot{e}_1)$$
$$+ (2L_2^i l_3^i S_{12} S_{123} + 2L_2^i l_3^i C_{12} C_{123} + L_1^i L_2^i S_1 S_{12} + L_1^i L_2^i C_1 C_{12}$$
$$+ L_1^i l_3^i S_1 S_{123} + L_1^i l_3^i C_1 C_{123} + L_2^i L_2^i + l_3^i l_3^i)(\ddot{q}_{d2} + \lambda_2 \dot{e}_2)$$
$$+ (L_1^i l_3^i S_1 S_{123} + L_1^i l_3^i C_1 C_{123} + L_2^i l_3^i S_{12} S_{123}$$
$$+ L_2^i l_3^i C_{12} C_{123} + l_3^i l_3^i)(\ddot{q}_{d3} + \lambda_3 \dot{e}_3)$$
$$+ g L_1 C_1 + g L_2 C_{12} + g l_3 C_{123}$$
$$+ (L_1^i L_2^i S_1 C_{12} - L_1^i L_2^i C_1 S_{12})\dot{q}_1(\dot{q}_{d2} + \lambda_2 e_2)$$
$$+ (L_1^i L_2^i S_1 C_{12} - L_1^i L_2^i C_1 S_{12})\dot{q}_2(\dot{q}_{d1} + \lambda_1 e_1)$$
$$+ (L_1^i l_3^i S_1 C_{123} - L_1^i l_3^i C_1 S_{123})\dot{q}_1(\dot{q}_{d2} + \lambda_2 e_2)$$
$$+ (L_1^i l_3^i S_1 C_{123} - L_1^i l_3^i C_1 S_{123})\dot{q}_2(\dot{q}_{d1} + \lambda_1 e_1)$$
$$+ (L_1^i l_3^i S_1 C_{123} - L_1^i l_3^i C_1 S_{123})\dot{q}_1(\dot{q}_{d3} + \lambda_3 e_3)$$
$$+ (L_1^i l_3^i S_1 C_{123} - L_1^i l_3^i C_1 S_{123})\dot{q}_3(\dot{q}_{d1} + \lambda_1 e_1)$$
$$+ (L_2^i l_3^i S_{12} C_{123} - L_2^i l_3^i C_{12} S_{123})\dot{q}_1(\dot{q}_{d3} + \lambda_3 e_3)$$
$$+ (L_2^i l_3^i S_{12} C_{123} - L_2^i l_3^i C_{12} S_{123})\dot{q}_3(\dot{q}_{d1} + \lambda_1 e_1)$$
$$+ (L_1^i l_3^i S_1 C_{123} - L_1^i l_3^i C_1 S_{123})\dot{q}_2(\dot{q}_{d3} + \lambda_3 e_3)$$
$$+ (L_1^i l_3^i S_1 C_{123} - L_1^i l_3^i C_1 S_{123})\dot{q}_3(\dot{q}_{d2} + \lambda_2 e_2)$$
$$+ (L_2^i l_3^i S_{12} C_{123} - L_2^i l_3^i C_{12} S_{123})\dot{q}_2(\dot{q}_{d3} + \lambda_3 e_3)$$
$$+ (L_2^i l_3^i S_{12} C_{123} - L_2^i l_3^i C_{12} S_{123})\dot{q}_3(\dot{q}_{d2} + \lambda_2 e_2)$$
$$+ (L_1^i L_2^i S_1 C_{12} - L_1^i L_2^i C_1 S_{12})\dot{q}_2(\dot{q}_{d2} + \lambda_2 e_2)$$
$$+ (L_1^i l_3^i S_1 C_{123} - L_1^i l_3^i C_1 S_{123})\dot{q}_2(\dot{q}_{d2} + \lambda_2 e_2)$$
$$+ (L_1^i l_3^i S_1 C_{123} - L_1^i l_3^i C_1 S_{123})\dot{q}_3(\dot{q}_{d3} + \lambda_3 e_3)$$
$$+ (L_2^i l_3^i S_{12} C_{123} - L_2^i l_3^i C_{12} S_{123})\dot{q}_3(\dot{q}_{d3} + \lambda_3 e_3),$$

$$Y_{14}^i = \ddot{q}_{d1} + \lambda_1 \dot{e}_1,$$

$$Y_{15}^i = \ddot{q}_{d1} + \lambda_1 \dot{e}_1 + \ddot{q}_{d2} + \lambda_2 \dot{e}_2,$$

$$Y_{16}^i = \ddot{q}_{d1} + \lambda_1 \dot{e}_1 + \ddot{q}_{d2} + \lambda_2 \dot{e}_2 + \ddot{q}_{d3} + \lambda_3 \dot{e}_3,$$

$$Y_{21}^i = 0,$$

$$Y_{22}^i = (L_1^i l_2^i S_1 S_{12} + L_1^i l_2^i C_1 C_{12} + l_2^i l_2^i)(\ddot{q}_{d1} + \lambda_1 \dot{e}_1)$$
$$+ l_2^i l_2^i(\ddot{q}_{d2} + \lambda_2 \dot{e}_2) + g l_2 C_{12}$$
$$+ (L_1^i l_2^i S_1 C_{12} - L_1^i l_2^i C_1 S_{12})\dot{q}_2(\dot{q}_{d1} + \lambda_1 e_1)$$
$$+ (L_1^i l_2^i C_1 S_{12} - L_1^i l_2^i S_1 C_{12})\dot{q}_1(\dot{q}_{d1} + \lambda_1 e_1),$$

$$Y_{23}^i = (2L_2^i l_3^i S_{12} S_{123} + 2L_2^i l_3^i C_{12} C_{123} + L_1^i L_2^i S_1 S_{12} + L_1^i L_2^i C_1 C_{12}$$
$$+L_1^i l_3^i S_1 S_{123} + L_1^i l_3^i C_1 C_{123} + L_2^i L_2^i + l_3^i l_3^i)(\ddot{q}_{d1} + \lambda_1 \dot{e}_1)$$
$$+(2L_2^i l_3^i S_{12} S_{123} + 2L_2^i l_3^i C_{12} C_{123} + L_2^i L_2^i + l_3^i l_3^i)(\ddot{q}_{d2} + \lambda_2 \dot{e}_2)$$
$$+(L_2^i l_3^i S_{12} S_{123} + L_2^i l_3^i C_{12} C_{123} + l_3^i l_3^i)(\ddot{q}_{d3} + \lambda_3 \dot{e}_3)$$
$$+gL_2 C_{12} + gl_3 C_{123} + (L_1^i L_2^i S_1 C_{12} - L_1^i L_2^i C_1 S_{12})\dot{q}_1(\dot{q}_{d2} + \lambda_2 e_2)$$
$$+(L_1^i l_3^i S_1 C_{123} - L_1^i l_3^i C_1 S_{123})\dot{q}_1(\dot{q}_{d2} + \lambda_2 e_2)$$
$$+(L_2^i l_3^i S_{12} C_{123} - L_2^i l_3^i C_{12} S_{123})\dot{q}_1(\dot{q}_{d3} + \lambda_3 e_3)$$
$$+(L_2^i l_3^i S_{12} C_{123} - L_2^i l_3^i C_{12} S_{123})\dot{q}_3(\dot{q}_{d1} + \lambda_1 e_1)$$
$$+(L_2^i l_3^i S_{12} C_{123} - L_2^i l_3^i C_{12} S_{123})\dot{q}_2(\dot{q}_{d3} + \lambda_3 e_3)$$
$$+(L_2^i l_3^i S_{12} C_{123} - L_2^i l_3^i C_{12} S_{123})\dot{q}_3(\dot{q}_{d2} + \lambda_2 e_2)$$
$$+(L_1^i L_2^i C_1 S_{12} - L_1^i L_2^i S_1 C_{12})\dot{q}_1(\dot{q}_{d1} + \lambda_1 e_1)$$
$$+(L_1^i l_3^i C_1 S_{123} - L_1^i l_3^i S_1 C_{123})\dot{q}_1(\dot{q}_{d1} + \lambda_1 e_1)$$
$$+(L_2^i l_3^i S_{12} C_{123} - L_2^i l_3^i C_{12} S_{123})\dot{q}_3(\dot{q}_{d3} + \lambda_3 e_3),$$

$$Y_{24}^i = 0,$$
$$Y_{25}^i = \ddot{q}_{d1} + \lambda_1 \dot{e}_1 + \ddot{q}_{d2} + \lambda_2 \dot{e}_2,$$
$$Y_{26}^i = \ddot{q}_{d1} + \lambda_1 \dot{e}_1 + \ddot{q}_{d2} + \lambda_2 \dot{e}_2 + \ddot{q}_{d3} + \lambda_3 \dot{e}_3,$$
$$Y_{31}^i = 0, \ Y_{32}^i = 0,$$
$$Y_{33}^i = (L_1^i l_3^i S_1 S_{123} + L_1^i l_3^i C_1 C_{123} + L_2^i l_3^i S_{12} S_{123}$$
$$+L_2^i l_3^i C_{12} C_{123} + l_3^i l_3^i)(\ddot{q}_{d1} + \lambda_1 \dot{e}_1)$$
$$+(L_2^i l_3^i S_{12} S_{123} + L_2^i l_3^i C_{12} C_{123} + l_3^i l_3^i)(\ddot{q}_{d2} + \lambda_2 \dot{e}_2)$$
$$+l_3^i l_3^i(\ddot{q}_{d3} + \lambda_3 \dot{e}_3) + gl_3 C_{123}$$
$$+(L_2^i l_3^i C_{12} S_{123} - L_2^i l_3^i S_{12} C_{123})\dot{q}_1(\dot{q}_{d2} + \lambda_2 e_2)$$
$$+(L_2^i l_3^i C_{12} S_{123} - L_2^i l_3^i S_{12} C_{123})\dot{q}_2(\dot{q}_{d1} + \lambda_1 e_1)$$
$$+(L_1^i l_3^i S_1 C_{123} - L_1^i l_3^i C_1 S_{123})\dot{q}_1(\dot{q}_{d3} + \lambda_3 e_3)$$
$$+(L_2^i l_3^i S_{12} C_{123} - L_2^i l_3^i C_{12} S_{123})\dot{q}_1(\dot{q}_{d3} + \lambda_3 e_3)$$
$$+(L_2^i l_3^i S_{12} C_{123} - L_2^i l_3^i C_{12} S_{123})\dot{q}_2(\dot{q}_{d3} + \lambda_3 e_3)$$
$$+(L_1^i l_3^i C_1 S_{123} - L_1^i l_3^i S_1 C_{123})\dot{q}_1(\dot{q}_{d1} + \lambda_1 e_1)$$
$$+(L_2^i l_3^i C_{12} S_{123} - L_2^i l_3^i S_{12} C_{123})\dot{q}_1(\dot{q}_{d1} + \lambda_1 e_1)$$
$$+(L_2^i l_3^i C_{12} S_{123} - L_2^i l_3^i S_{12} C_{123})\dot{q}_2(\dot{q}_{d2} + \lambda_2 e_2),$$

$$Y_{34}^i = 0,$$
$$Y_{35}^i = 0,$$
$$Y_{36}^i = \ddot{q}_{d1} + \lambda_1 \dot{e}_1 + \ddot{q}_{d2} + \lambda_2 \dot{e}_2 + \ddot{q}_{d3} + \lambda_3 \dot{e}_3,$$
$$C_1 = \cos(q_1),$$
$$C_{12} = \cos(q_1 + q_2),$$
$$C_{123} = \cos(q_1 + q_2 + q_3),$$
$$S_1 = \sin(q_1),$$
$$S_{12} = \sin(q_1 + q_2),$$
$$S_{123} = \sin(q_1 + q_2 + q_3).$$

Bibliography

[1] M. J. Marquez. *Nonlinear Control Systems: Analysis and Design.* Wiley-Interscience, Hoboken, New Jersey, 2003.

[2] F. L. Lewis, D. M. Dawson, and C. T. Abdallah. *Robot Manipulators Control: Second Edition, Revised and Expanded.* Marcel Dekker, Inc., New York, USA, 2004.

[3] F. L. Lewis, S. Jagannathan, and A. Yesildirak. *Neural Network Control of Robot Manipulators and Non-Linear Systems.* CRC, 1998.

[4] R. Kelly, V. Santibanez, and A. Loria. *Control of Robot Manipulators in Joint Space.* Springer, New York, USA, 2005.

[5] C.-H. Chen, D. S. Naidu, A. Perez-Gracia, and M. P. Schoen. "A hybrid control strategy for five-fingered smart prosthetic hand," in *Joint 48th IEEE Conference on Decision and Control (CDC) and 28th Chinese Control Conference (CCC),* pp. 5102–5107, Shanghai, P. R. China, December 16–18, 2009.

[6] D. S. Naidu. *Optimal Control Systems.* CRC Press, a Division of Taylor & Francis, Boca Raton, FL and London, UK, 2003 (A vastly expanded and updated version of this book, is under preparation for publication in 2017).

[7] C.-H. Chen, D. S. Naidu, A. Perez-Gracia, and M. P. Schoen. "A hybrid optimal control strategy for a smart prosthetic hand," in *Proceedings of the ASME 2009 Dynamic Systems and Control Conference (DSCC),* Hollywood, California, USA, October 12–14, 2009 (No. DSCC2009–2507).

[8] C.-H. Chen and D. S. Naidu. "Hybrid genetic algorithm PID control for a five-fingered smart prosthetic hand," in *Proceedings of the Sixth International Conference on Circuits, Systems and Signals (CSS'11),* pp. 57–63, Vouliagmeni Beach, Athens, Greece, March 7–9, 2012.

[9] C.-H. Chen and D. S. Naidu. A modified optimal control strategy for a five-finger robotic hand. *International Journal of Robotics and Automation Technology,* 1(1):3–10, November 2014.

[10] F. L. Lewis, S. Jagannathan, and A. Yesildirek. *Neural Network Control of Robotic Manipulators and Nonlinear Systems.* Taylor & Francis, London, UK, 1999.

[11] C.-H. Chen, D. S. Naidu, and M. P. Schoen. Adaptive control for a five-fingered prosthetic hand with unknown mass and inertia. *World Scientific and Engineering Academy and Society (WSEAS) Journal on Systems,* 10(5):148–161, May 2011.

[12] S. Arimoto. *Control Theory of Multi-fingered Hands: A Modeling and Analytical–Mechanics Approach for Dexterity and Intelligence.* Springer-Verlag, London, UK, 2008.

[13] C.-H. Chen, D. S. Naidu, A. Perez-Gracia, and M. P. Schoen. "A hybrid adaptive control strategy for a smart prosthetic hand," in *The 31st Annual International Conference of the IEEE Engineering Medicine and Biology Society (EMBS)*, pp. 5056–5059, Minneapolis, Minnesota, USA, September 2–6, 2009.

[14] C.-H. Chen, K. W. Bosworth, M. P. Schoen, S. E. Bearden, D. S. Naidu, and A. Perez-Gracia. "A study of particle swarm optimization on leukocyte adhesion molecules and control strategies for smart prosthetic hand," in *2008 IEEE Swarm Intelligence Symposium (IEEE SIS08)*, St. Louis, Missouri, USA, September 21–23, 2008.

[15] C.-H. Chen, D. S. Naidu, A. Perez-Gracia, and M. P. Schoen. "Fusion of hard and soft control techniques for prosthetic hand," in *Proceedings of the International Association of Science and Technology for Development (IASTED) International Conference on Intelligent Systems and Control (ISC 2008)*, pp. 120–125, Orlando, Florida, USA, November 16–18, 2008.

[16] C.-H. Chen, D. S. Naidu, and M. P. Schoen. "An adaptive control strategy for a five-fingered prosthetic hand," in *The 14th World Scientific and Engineering Academy and Society (WSEAS) International Conference on Systems, Latest Trends on Systems (Volume II)*, pp. 405–410, Corfu Island, Greece, July 22–24, 2010.

[17] C.-H. Chen and D. S. Naidu. Hybrid control strategies for a five-finger robotic hand. *Biomedical Signal Processing and Control*, 8(4):382–390, July 2013.

CHAPTER 6

FUSION OF HARD AND SOFT CONTROL STRATEGIES II

In this chapter we present the fusion, hybrid or integration of hard control (HC) and soft control (SC) strategies to enhance the performance in terms of controlling a robotic/prosthetic hand that could not be achieved by using either HC or SC strategy. In Section 6.1, we introduce fuzzy-logic-based proportional-derivative (PD) fusion control strategy and in Section 6.2 we present genetic-algorithm-based proportional-integral-derivative (PID) fusion control strategy.

6.1 Fuzzy Logic Based PD Fusion Control Strategy

Figure 6.1 shows the block diagram of the fusion FL-based PD controller for the proposed five-finger robotic hand with control input signal as

$$\mathbf{u}(t) = -\mathbf{K_P}(t)\mathbf{e}(t) - \mathbf{K_D}(t)\dot{\mathbf{e}}(t) \qquad (6.1.1)$$

and the proportional $\mathbf{K_P}(t)$ and derivative $\mathbf{K_D}(t)$ diagonal gain matrices with time-varying t. We rewrite (5.1.17) as

$$\boldsymbol{\tau}(t) \quad = \quad \mathbf{M}(\mathbf{q}(t))\left[\ddot{\mathbf{q}}_d(t) + \mathbf{K_P}(t)\mathbf{e}(t) + \mathbf{K_D}(t)\dot{\mathbf{e}}(t) + \mathbf{N}(\mathbf{q}(t), \dot{\mathbf{q}}(t))\right]. \quad (6.1.2)$$

Fusion of Hard and Soft Control Strategies for the Robotic Hand, By C.-H. Chen and D. S. Naidu **203**
© 2017 by the Institute of Electrical and Electronic Engineers, Inc. Published 2017 by John Wiley & Sons, Inc.

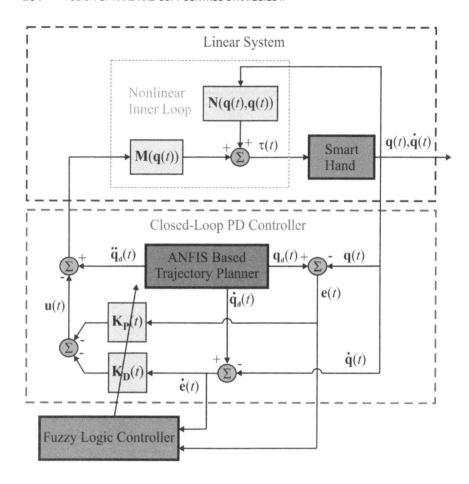

Figure 6.1 Block Diagram of the Proposed Fusion Fuzzy-Logic-Based Proportional-Derivative Controller for a Five-Finger Robotic Hand

Then we use Mamdani-type fuzzy inference system to tune the time-varying parameters $K_P(t)$ and $K_D(t)$ of the closed-loop PD controller due to its simple *min–max* structure [1]. It is noteworthy that S. J. Ovaska compared 12 intelligent fusion systems with HC and SC in structural fusion categories and gave brief definitions of the used fusion grade, which is qualitative measure to describe the strength of a specific connection between HC and SC structures: low, moderate, high, and very high [2]. In his structural fusion categories, the fusion grade of the current proposed fusion FL-based PD controller is classified as *very high*.

Mamdani model was proposed by Mamdani and Assilian to control a steam engine and boiler combination using a set of linguistic control rules obtained from experienced human operators in 1975 [1]. Since that time, the Mamdani system has become the commonly used fuzzy inference approach because of its simple *min–max* structure.

Figure 6.2(a) shows the block diagram of Mamdani-type fuzzy logic controller (FLC). The crisp inputs (errors $e(t)$ and error rates $\dot{e}(t)$) from the robotic hand are fuzzified by seven triangular membership functions as shown in Figure 6.2(b). The number of membership functions is determined by trials and errors. If the number is below seven, the model output will not satisfactorily follow the output of the robotic hand. Then the fuzzy inputs are parallel processed in fuzzy system by human knowledge reasoning and 7×7 logic "IF-THEN" rules, which are listed in Table 6.1. For instance, "IF" (the error $e(t)$ is negative small, **NS**) AND (the error rate $\dot{e}(t)$ is positive medium, **PM**), "THEN" ($\mathbf{K_P}(t)$ is medium large, **ML**). The fuzzy output is then defuzzified by another seven triangular membership functions (Figure 6.2(b)) to generate crisp output $\mathbf{K_P}(t)$. Figure 6.2(c) shows the output surface $\mathbf{K_P}(t)$ of this fuzzy inference system. Similarly, the other crisp output $\mathbf{K_D}(t)$ is computed by the same procedure. The adaptive $\mathbf{K_P}(t)$ and $\mathbf{K_D}(t)$ parameters are used in the closed-loop PD controller.

Table 6.1   A 7×7 Fuzzy Logic "IF-THEN" Rule Base

$\dot{e}\backslash e$	NL	NM	NS	ZR	PS	PM	PL
NL	ZR	ZR	ZR	ZR	VS	S	SM
NM	ZR	ZR	ZR	VS	S	SM	ML
NS	ZR	ZR	VS	S	SM	ML	L
ZR	ZR	VS	S	SM	ML	L	VL
PS	VS	S	SM	ML	L	VL	VL
PM	S	SM	ML	L	VL	VL	VL
PL	SM	ML	L	VL	VL	VL	VL

N: negative; P: positive; ZR: zero; L: large;
M: medium; S: small; V: very

Summarizing, errors $e(t)$ and error changes $\dot{e}(t)$ are calculated by actual angles $q(t)$ and desired angles $q_d(t)$, which are computed by adaptive neuro-fuzzy inference system (ANFIS) trajectory planner. Then FLC tunes all time-varying parameters $\mathbf{K_P}(t)$ and $\mathbf{K_D}(t)$ of the closed-loop PD control so that the required torques $\tau(t)$ of the robotic hand nonlinear system is computed by control input signal $u(t)$.

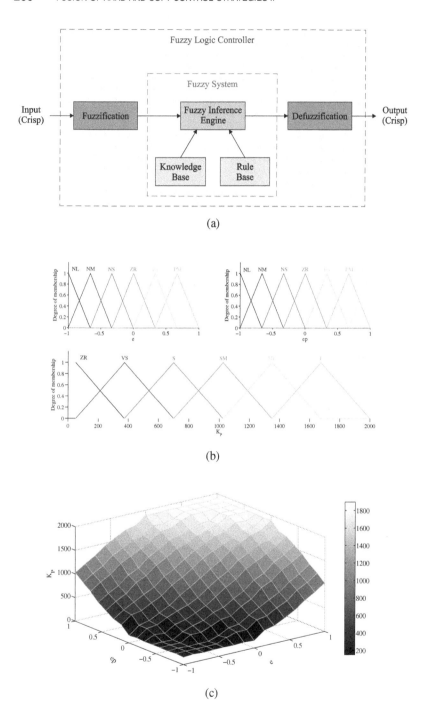

Figure 6.2 Structure of Fuzzy Logic Controller (FLC): (a) Block Diagram of Mamdani-Type FLC. (b) All Membership Functions for Two Inputs (Upper) and One Output (Lower) Using 7 Triangular Membership Functions. (c) The Output Surface of Fuzzy Inference System Based on Two Crisp Inputs and 49 Fuzzy Logic Knowledge Rules.

6.1.1 Simulation Results and Discussion

To compare the precision and energy cost of HC, SC, and the proposed fusion control strategies, we present simulations with PD and PID controllers and fuzzy inference system tuned PD controller for a five-finger robotic hand with 14-DOF to grasp a rectangular rod as shown in Figure 2.11. The parameters of the two-link thumb/three-link fingers are related to desired trajectory [3]. All parameters of the robotic hand selected for the simulations are given in Table 5.3 and the side length and length of the target rectangular rod are 0.010 and 0.100 (m), respectively. The relating parameters between the global coordinate and the local coordinates are defined in Figure 2.12 and the values are given in Table 5.4. When the thumb and other four fingers are performing extension/flexion movements, the workspace of fingertips is restricted to the maximum angles of joints. Referring to inverse kinematics, the first and second joint angles of the thumb are constrained within the ranges of $[0, 90]$ and $[-80, 0]$ (deg). The first, second, and third joint angles of the other four fingers are constrained in the ranges of $[0, 90]$, $[0, 110]$, and $[0, 80]$ (deg), respectively [4].

Moreover, each link of all fingers is assumed to be a circular cylinder with a radius ($R = 0.010$ m), so the inertia I_{zzk}^j of each link k ($=1$–3) of all fingers j ($= t, i, m, r,$ and l) is calculated as

$$I_{zzk}^j = \frac{1}{4}m_k^j R^2 + \frac{1}{3}m_k^j L_k^{j\,2}. \tag{6.1.3}$$

All initial values of actual angles are zero and the diagonal coefficients, $\mathbf{K_P}(t)$, $\mathbf{K_I}(t)$, and $\mathbf{K_D}(t)$, for the PD and PID controllers alone are arbitrarily chosen as 100. From the derived dynamic and control models, the control signal $\mathbf{u}(t)$ and torque $\boldsymbol{\tau}(t)$ are computed after the parameters ($\mathbf{K_P}(t)$ and $\mathbf{K_D}(t)$) are selected. Figure 6.3 shows tracking errors $e_1^t(t)$ and $e_2^t(t)$ (left column) and desired/actual angles $q_1^t(t)$ and $q_2^t(t)$ (right column) of joints 1 (top row) and 2 (bottom row) for two-link thumb using PD (dash line), PID (dot line), FL (short dash line), and fusion of FL and PD (solid line) controllers. The tracking errors for both PD and FL controllers are convergent to zero within 5 seconds without overshooting, but PID controller takes longer (approximately 10 seconds) with overshooting and oscillation. The proposed fusion control of FL and PD using parameters $\mathbf{K_P}(t) \in [50, 1000]$ and $\mathbf{K_D}(t) \in [50, 500]$ provides 5–10-fold faster convergence than PD, PID, and FLC alone. FLC includes two inputs (errors and error rates) and one output (control input signals). To further study whether the parameter range influences tracking errors, we found that the larger the parameter range, the faster the convergent speed for the range $\mathbf{K_P}(t) \in [50, 2000]$ without additional computational time [5]. These results clearly demonstrate that fusion of SC and HC is superior to either HC or SC methodology alone.

The time-varying computed control signals ($u_1^t(t)$ and $u_2^t(t)$) and torques ($\tau_1^t(t)$ and $\tau_2^t(t)$) for two-link thumb are shown in Figure 6.4, suggesting that the presented fusion FL-PD controller requires more torques (energy cost) than PD, PID, and FL controllers in order to obtain faster convergent tracking errors.

For the three-link index finger, Figure 6.5 shows tracking errors and desired/actual angles and Figure 6.6 shows control signals and torques, respectively. Similar results

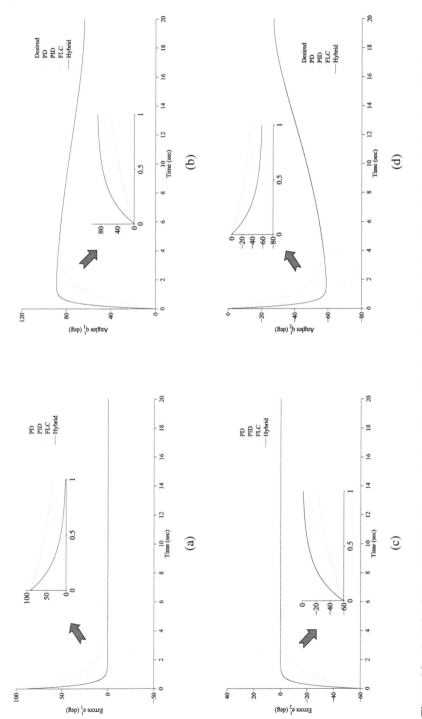

Figure 6.3 Tracking Errors and Desired/Actual Angles of Joints 1 (a,b) and 2 (c,d) for Two-Link Thumb Using PD (dash line), PID (dot line), FL (short dash line), and Hybrid/Fusion of FL and PD (solid line) Controllers

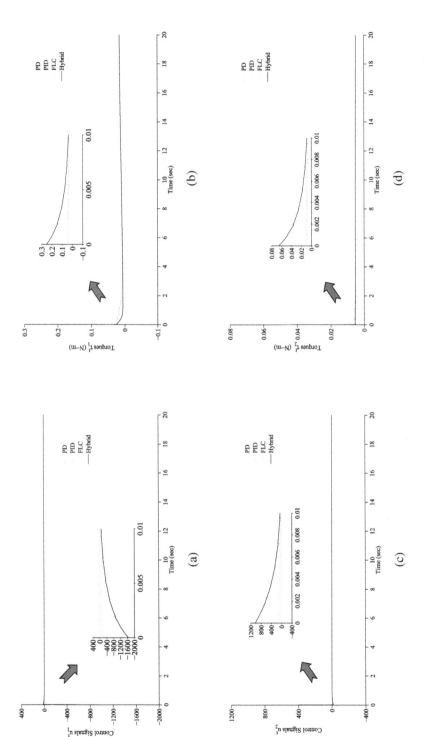

Figure 6.4   Control Signals and Actuated Torques of Joints 1 (a,b) and 2 (c,d) for Two-Link Thumb Using PD (dash line), PID (dot line), FL (short dash line), and Hybrid/Fusion of FL and PD (solid line) Controllers

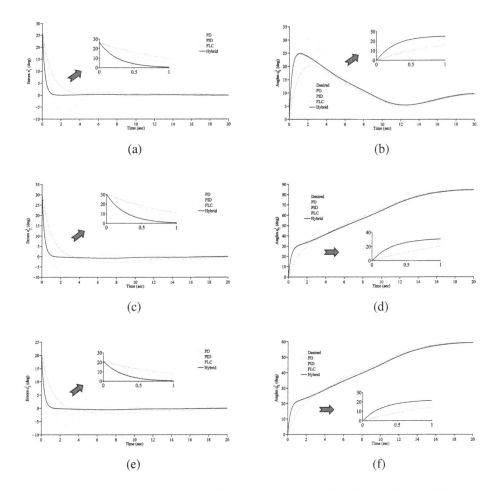

Figure 6.5   Tracking Errors and Desired/Actual Angles of Joints 1 (a,b), 2 (c,d), and 3 (e,f) for Three-Link Index Finger Using PD (dash line), PID (dot line), FL (short dash line), and Hybrid/Fusion of FL and PD (solid line) Controllers

of tracking errors for the remaining three-link fingers (middle, ring, and little) are consistently obtained (data not shown). As increasing DOF, PID control still shows undesirable features of overshooting and oscillation; both PD and FL controllers reduce accuracy and convergent speed. As for the fusion FL-based PD control, it increases required power, but this fusion controller still holds fast convergence and high accuracy as the DOF increases. These findings suggest that improving accuracy, the proposed fusion control strategies are potentially applied to industrial robotic applications for hazardous environments, surgery, etc., but this fusion approach has the limitation of clinical applications to robotic devices due to the current battery capac-

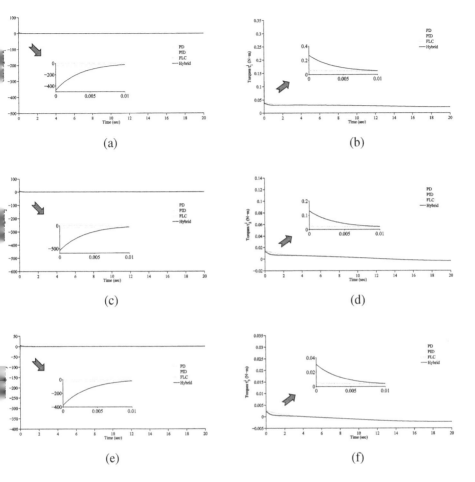

Figure 6.6   Control Signals and Actuated Torques of Joints 1 (a,b), 2 (c,d), and 3 (e,f) for Three-Link Index Finger Using PD (dash line), PID (dot line), FL (short dash line), and Hybrid/Fusion of FL and PD (solid line) Controllers

ity. This limitation may be solved by implementing HC and/or SC methodology to optimize control inputs embedded in cost function (performance index) [6].

6.2   Genetic Algorithm Based PID Fusion Control Strategy

Figure 6.7 shows the block diagram of a fusion GA-based PID controller with control

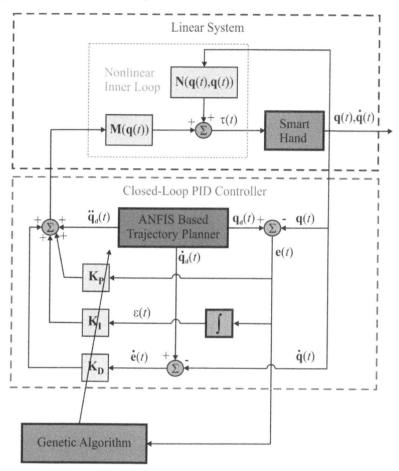

Figure 6.7   Block Diagram of the Fusion GA-Based PID Controller for the 14-DOF, Five-Finger Robotic Hand

signal as

$$u(t) = -\mathbf{K_P}e(t) - \mathbf{K_I} \int e(t)dt - \mathbf{K_D}\dot{e}(t) \qquad (6.2.1)$$

with the proportional $\mathbf{K_P}$, integral $\mathbf{K_I}$, and derivative $\mathbf{K_D}$ diagonal gain matrices. We then rewrite (5.1.17) as

$$\boldsymbol{\tau}(t) = \mathbf{M(q}(t))[\ddot{\mathbf{q}}_d(t) + \mathbf{K_P}e(t) + \mathbf{K_I} \int e(t)dt$$
$$+\mathbf{K_D}\dot{e}(t)) + \mathbf{N(q}(t), \dot{\mathbf{q}}(t)]. \qquad (6.2.2)$$

Then we use GA to tune all gain coefficients $\mathbf{K_P}$, $\mathbf{K_I}$, and $\mathbf{K_D}$ of PID controller. From the derived dynamic and control models, after the parameters ($\mathbf{K_P}$, $\mathbf{K_I}$, and $\mathbf{K_D}$) are determined, the torque matrix τ is calculated, and then the squared-tracking errors $e_i^j(t)$ of the joint i of the finger j are obtained. Thus, the total error $E(t)$, a time-dependent function, is defined as

$$E(t) = \int_{t_0}^{t_f} (e_i^j(t))^2 dt, \qquad (6.2.3)$$

where t_0 and t_f are initial and terminal time, respectively. As shown in Figure 4.7, the tuned diagonal parameters ($\mathbf{K_P}$, $\mathbf{K_I}$, and $\mathbf{K_D}$) and the total error $E(t)$ of PID controller by GA are obtained and listed in Table 6.2 based on our previous study [7].

Table 6.2   Parameter Selection of GA-Tuned PID Controller and Computed Total Errors

	Input			Output
Fingers	$\mathbf{K_P}$	$\mathbf{K_I}$	$\mathbf{K_D}$	$E(t)$
Case I	[976,956]	[779,279]	[170,236]	0.3107
Case II	[988,999]	[78,848]	[80,109]	0.1557
Case III	[199,198]	[127,157]	[104,102]	0.8100
Index	[794,398,960]	[960,918,914]	[15,59,242]	0.0465
Middle	[794,398,960]	[960,918,914]	[15,59,242]	0.1003
Ring	[794,398,960]	[960,918,914]	[15,59,242]	0.0465
Little	[794,398,960]	[960,918,914]	[15,59,242]	0.0607

6.2.1 Simulation Results and Discussion

To study whether the tuned parameter range influences total tracking errors, we design three different cases with altering lower and upper bounds of tuned parameter ranges for two-link thumb. Cases I, II, and III for the thumb represent that the PID parameters $\mathbf{K_P}$, $\mathbf{K_I}$, and $\mathbf{K_D}$ are constricted in three different bounded ranges [100,1000], [50,1000], and [100,200], respectively. Figures 6.8 and 6.9 show the tracking errors and desired/actual angles of joints 1 and 2 of PID and GA-based PID controllers for thumb. These simulations show that the large ranges [100,1000] (Case I) and [50,1000] (Case II) provide better results than the PID controller parameters arbitrarily chosen as 100. However, the small range [100,200] (Case III) gives worse result than the PID controller alone. These results suggest that the bigger parameter range, the smaller the total error. Cases I and II explain that GA finds some parameter values $\in$ [100,1000] and [50,100] escaping the local minimum area. Case III covers the value 100 in lower bound, but both total error and convergent speed are even worse than PID alone, suggesting that GA performs better for a large range, but is poor for searching on the boundary. To further consider the convergent speed, Case I gives smaller total error, but does not improve its convergent speed when comparing to PID control alone. Yet, Case II gives good total error and convergent speed. Case III gives poor total error and convergent speed. Taken together, these results imply that the global minimum could be located in the ranges [50,100] and [200,1000] and the parameter ranges play an important role in GA tuning. Based on these findings, we use the range [50,1000] for the remaining three-link fingers. Figures 6.10–6.13 show the simulations of PID and GA-based PID controllers for the remaining three-link fingers [8].

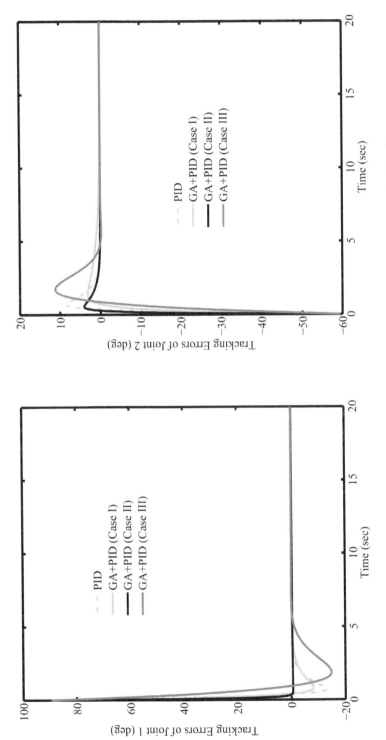

Figure 6.8   Tracking Errors of Joints 1 and 2 of PID and GA-Based PID Controllers for Thumb

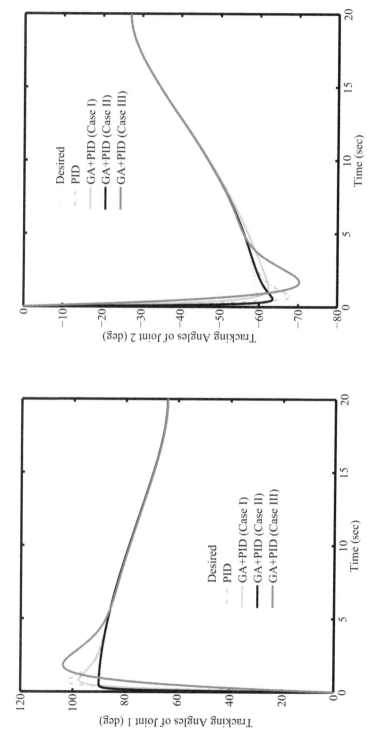

Figure 6.9   Tracking Angles of Joints 1 and 2 of PID and GA-Based PID Controllers for Thumb

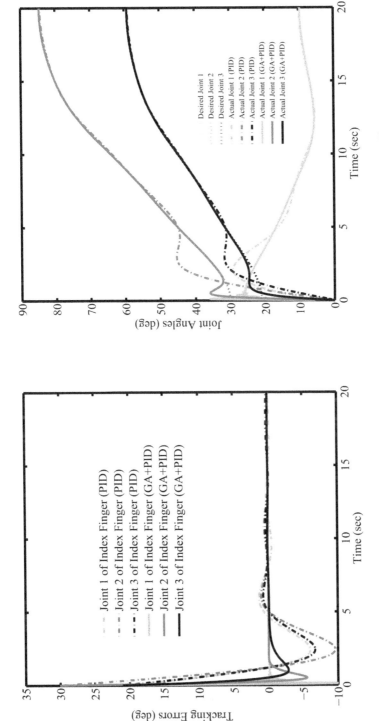

Figure 6.10   Tracking Errors and Joint Angles of PID and GA-Based PID Controllers for Index Finger

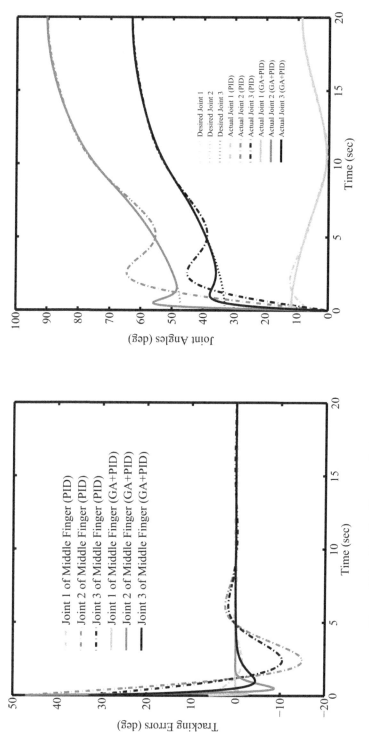

Figure 6.11   Tracking Errors and Joint Angles of PID and GA-Based PID Controllers for Middle Finger

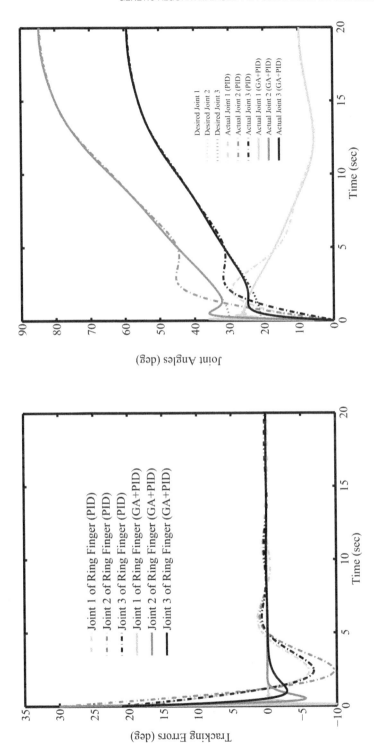

Figure 6.12 Tracking Errors and Joint Angles of PID and GA-Based PID Controllers for Ring Finger

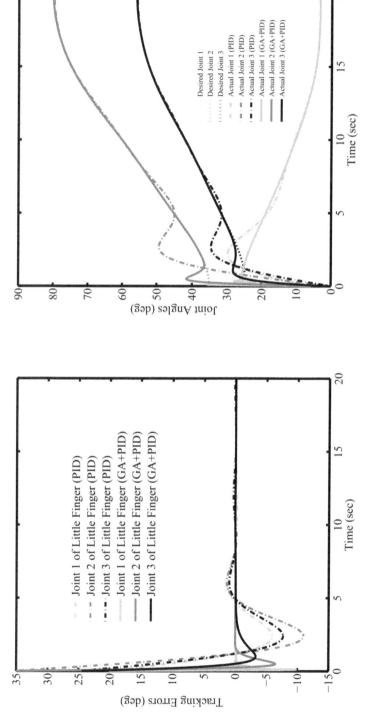

Figure 6.13 Tracking Errors and Joint Angles of PID and GA-Based PID Controllers for Little Finger

Bibliography

[1] E. H. Mamdani and S. Assilian. An experiment in linguistic synthesis with a fuzzy logic controller. *International Journal of Man-Machine Studies*, 7(1):1–13, 1975.

[2] S. J. Ovaska, H. F. VanLandingham, and A. Kamiya. Fusion of soft computing and hard computing in industrial applications: An overview. *IEEE Transactions on Systems, Man, and Cybernetics, Part C: Applications and Reviews*, 32(2):72–79, May 2002.

[3] S. Arimoto. *Control Theory of Multi-fingered Hands: A Modeling and Analytical–Mechanics Approach for Dexterity and Intelligence*. Springer-Verlag, London, UK, 2008.

[4] P. K. Lavangie and C. C. Norkin. *Joint Structure and Function: A Comprehensive Analysis, Third Edition*. F. A. Davis Company, Philadelphia, Pennsylvania, USA, 2001.

[5] C.-H. Chen and D. S. Naidu. "Fusion of fuzzy logic and PD control for a five-fingered smart prosthetic hand," in *Proceedings of the 2011 IEEE International Conference on Fuzzy Systems (FUZZ–IEEE 2011)*, pp. 2108–2115, Taipei, Taiwan, June 27–30, 2011.

[6] C.-H. Chen and D. S. Naidu. Hybrid control strategies for a five-finger robotic hand. *Biomedical Signal Processing and Control*, 8(4):382–390, July 2013.

[7] C.-H. Chen and D. S. Naidu. "Hybrid genetic algorithm PID control for a five-fingered smart prosthetic hand," in *Proceedings of the Sixth International Conference on Circuits, Systems and Signals (CSS'11)*, pp. 57–63, Vouliagmeni Beach, Athens, Greece, March 7–9, 2012.

[8] C.-H. Chen and D. S. Naidu. A modified optimal control strategy for a five-finger robotic hand. *International Journal of Robotics and Automation Technology*, 1(1):3–10, November 2014.

CHAPTER 7

CONCLUSIONS AND FUTURE WORK

7.1 Conclusions

This book describes the fusion of *hard* control strategies such as PID, optimal, adaptive, and *soft* control strategies such as adaptive neuro-fuzzy inference system (ANFIS), genetic algorithms (GA), particle swarm optimization (PSO), for a robotic/prosthetic hand.

Chapter 2 addressed the forward kinematics, inverse kinematics, and differential kinematics models of a serial n revolute–joint planar two-link thumb, and three-link index finger. The fingertip (end-effector) positions of each finger were derived by forward kinematics. The joint angles of each finger (*joint space*) were obtained from the known fingertip positions (*Cartesian space*) by using inverse kinematics. Then, the workspaces of the fingertip were successfully generated. The linear and angular velocities and accelerations of fingertips were obtained by differential kinematics; the joint angular velocities and joint angular accelerations of each finger were then derived from the linear and angular velocities and accelerations of fingertips using the *geometric Jacobian*. Two trajectory planning, cubic polynomial and Bézier curve, functions were derived. The results were successfully applied to 14 DOF, five-fingered robotic hand model.

Fusion of Hard and Soft Control Strategies for the Robotic Hand, By C.-H. Chen and D. S. Naidu   **223**
© 2017 by the Institute of Electrical and Electronic Engineers, Inc. Published 2017 by John Wiley & Sons, Inc.

In Chapter 3, using the mathematical model of the actuator by using direct current (DC) motor and mechanical gears, the dynamic equations of hand motion were successfully derived via Lagrangian approach for two-link thumb and three-link fingers.

Chapter 4 successfully developed soft computing (SC) or computational intelligence (CI) techniques, including fuzzy logic (FL), neural network (NN), AN-FIS, tabu search (TS), GA, PSO, developed adaptive particle swarm optimization (APSO), and condensed hybrid optimization (CHO) as described below.

1. **ANFIS and GA**: Using ANFIS and GA methods, we successfully solved the inverse kinematics problems of three-link fingers. The simulations showed that the GA method although gave a better solution (error $\approx 10^{-7}$), took more execution time whereas the ANFIS gave a good solution (error $\approx 10^{-4}$) with less time. Therefore, this work used ANFIS method to find the inverse kinematics of three-link fingers.

2. **PSO**: As for the PSO dynamics investigation, we concluded that a small amount of negative velocity is useful and produced better results. In other words, during the procedure, most particles are looking for the same searching directions (positive weights), but a few particles are looking for different searching spaces (negative weights). In addition, techniques A–E produced agreeable performance, but it appears that E is the most costly.

3. **APSO**: The proposed APSO with changed updating velocity direction produced improved results compared to the generic PSO. To compare the standard deviation, APSO with changed updating velocity direction also yielded more stable results than the generic PSO. Because the APSO technique is attempting to use secant plane information to help determine good search directions, and for very smooth problems, possessing nicely defined minima, one would expect better performance.

4. **PSO in Inflammatory Applications**: The results demonstrated the utility of PSO in generating predictions about the integrated effects of multiple selectin–ligand pairs. These predictions can then be used to generate testable hypotheses. Use of this system will speed up the understanding how expression and regulation of multiple selectins–ligands contributes to inflammation *in vivo*.

5. **CHO**: The simulation results showed that the proposed CHO algorithm combines the advantages of TS and PSO and obtains robust results. However, on the higher-dimensional problems, CHO showed a sensitive dependence on selecting the most promising area from the promising list. The constant radius of the chosen promising area played a key role. Therefore, this parameter selection can be studied in future work.

In Chapter 5, hard control techniques, including feedback linearization, PD/PI/PID, optimal, and adaptive controllers, were developed for the robotic hand described in the previous chapters. The numerical simulation results with a PID controller and a finite-time linear quadratic optimal controller for the 14 DOF five-fingered smart robotic hand with realistic data showed the superiority of the optimal controller in suppressing the overshoots and transients associated with the PID controller.

Chapter 6 successfully developed the fusion of HC and SC to produce fusion of hard and soft control strategies to take advantage of the desired features from HC and SC. The numerical simulations with realistic data for a robotic hand with two-link thumb and three-link fingers with hybrid control strategy including fusion of PD controller and FL and hybrid of PID controller and GA demonstrated that the integration of SC and HC methodologies is superior to using either HC or SC technique alone.

7.2 Future Directions

The future directions may focus to develop the models and advanced control strategies for five-finger hand with 14 and 22 degrees-of-freedom (DOFs) to achieve tasks such as touching, holding, grasping, as shown in Figure 7.1. Next, the control strategies in real-time environment are proposed.

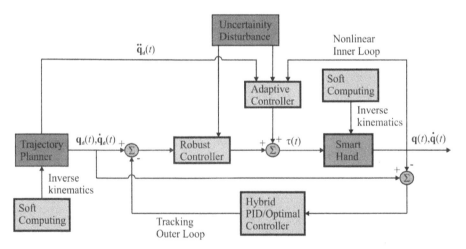

Figure 7.1   Block Diagram of Hybrid Control Strategies for Future Work

1. Develop an adaptive/robust controller for five-finger hand with 14 DOFs. The uncertainty and disturbance are considered in designing adaptive/robust controller including real-time environment.

2. Develop the PD/PI controller and adaptive/robust controller for the five-finger hand with 14 DOFs including the uncertainty factors including real-time environment.

3. Advanced exploration on soft computing techniques such as PSO, GA, to improve the computational cost including real-time environment.

4. Develop the optimal controller and adaptive/robust controller for five-finger hand with 14 DOFs including real-time environment.

5. Develop the hybrid optimal and adaptive/robust controller for five-finger hand with 14 DOFs using soft computing such as GA/PSO, to tune the parameters of the hybrid controller including real-time environment.

6. Develop the hybrid PD/PI/PID controller and adaptive/robust controller for five-finger hand with 14 DOFs using SC like GA/PSO, to tune the parameters of the hybrid controller including real-time environment.

7. Modify the hybrid control strategies for real-time application for the five-finger hand with 22 DOFs.

8. Integrate the hand motion, mechanical design, grasping manipulation planning, and EMG-based model for embedded hierarchical real-time implementation for five-finger hand with 22 DOFs.

During the last three decades, investigations have been carried out on the use of EMG signals to come up with a robotic hand to perform as many functions as possible. However, the use of EMG signals is limited in the number of possible human-like functions with as few electrodes as possible and also finally for the robotic hand to have a natural "cosmetic" appearance. Further, the EMG signal cannot provide any kind of feedback to the user [1]. One of the several possible solutions to overcome the limitations of surface-based EMG approach is neuroprosthesis. Here, we use an interface between the peripheral nervous system (PNS) and the "natural" neural interface to extract, record, and simulate the PNS in a selective way. Further, advanced in biocompatible neural interfaces can provide some sensory feedback to the user by stimulating the afferent nerves and allowing motor control of prosthesis leading to a "natural" EMG-based control. Based on this two possible ways of controlling robotic hand are the simple and non-intrusive EMG-based control and the more complicated, implantable EMG-based control.

As per the National Academies *Keck Futures Initiative*, "Smart Prosthetics: Exploring Assistive Devices for Body and Mind," as reported in [2], it was pointed out, "We can make smarter prostheses ... integrating engineering, medicine and social science." A related news item is on mind-controlled robotic arm ("$120 million man") showing the advances in neuroprosthetic technology at Johns Hopkins University [3].

Other interesting/promising ideas are in the area of three-dimensional printing as seen from [4] relating to "3-D printed prosthetics help kid athletes compete."

Bibliography

[1] M. Zecca, S. Micera, M. C. Carrozza, and P. Dario. Control of multifunctional prosthetic hands by processing the electromyographic signal. *Critical Reviews*^{TM} *in Biomedical Engineering*, 30:459–485, 2002 (Review article with 96 references).

[2] Arnold and Mabel Beckman Center of the National Academies. *NAKFI: Smart Prosthetics: Exploring Assistive Devices for Body and Mind: Task Group Summaries*, Irvine, California, USA, November 9–11, 2007.

[3] S. Grobart. This guy has a thought controlled robotic arm, November 2015.

[4] F. Imbert. 3-D printed prosthetics help kid athletes compete, November 2015.

Index

3-D printing, 226

actuator, 94

biomedical application, 149
Brunovsky canonical form, 162

Cartesian space, 66, 110, 223
chromosome, 118
Constraints
  controls, 167
  states, 167
control input, 162, 163
Coordinate
  global coordinate, 64, 177, 180
  local coordinate, 64, 177, 179, 180

degree of freedom, 1, 10, 20, 225
Denavit-Hartenberg, 58
dendrite, 108
differential Riccati equation, 168
direct current motor, 93
diversification, 114, 136
dynamic equation, 98, 99, 161, 224

electrical action potential, 108

electromyography, 5, 9–11, 19, 20, 226
end-effector, 49, 60, 72, 223

fuzzy if-then rule, 107, 110, 133
fuzzy set, 106, 108

gene, 118
global best position, 122

Hard control
  adaptive control, 11, 15, 19, 22, 170, 223
  feedback linearization, 22, 161, 224
  hierarchical control, 16
  optimal control, 16, 19, 22, 23, 167, 178, 223
  PD control, 10, 14, 15, 164, 203
  PI control, 165
  PID control, 17, 19, 22, 23, 165, 212, 223
  robust control, 16, 18, 22
homogeneous transformation, 50
human hand, 1, 6, 9, 48

inertia, 97, 98, 173, 175, 177
intensification, 116, 136, 145

EPILOGUE

Basis for the Proposed Book

The following are the publications (in chronological order without listing numerous quarterly reports submitted to the funding agency) by the authors and their colleagues forming the main basis for the proposed book:

1. J. C. K. Lai, M. P. Schoen, A. Perez-Gracia, D. S. Naidu, and S. W. Leung, "Prosthetic devices: Challenges and implications of robotic implants and biological interfaces," Special Issue on Micro and Nano Technologies in Medicine, Proceedings of the Institute of Mechanical Engineers (IMechE), London, UK, Part H, Journal of Engineering in Medicine, 221(2): 173–183, 2007. Listed as 1 of 20 in *Top 20 Articles, in the Domain of Article 17385571, Since its Publication (2007)* according to *BioMedLib: "Who is Publishing in My Domain?"* as on September 22, 2014 and as 1 of 20 as on March 17, 2015.

2. C.-H. Chen, K. W. Bosworth, and M. P. Schoen, "Investigation of particle swarm optimization dynamics," in *Proceedings of International Mechanical Engineering Congress and Exposition (IMECE) 2007*, Seattle, Washington, USA, November 11–15, 2007 (No. IMECE2007–41343).

3. C.-H. Chen, K. W. Bosworth, M. P. Schoen, S. E. Bearden, D. S. Naidu, and A. Perez-Gracia, "A study of particle swarm optimization on leukocyte adhesion molecules and control strategies for smart prosthetic hand," in *2008 IEEE Swarm Intelligence Symposium (IEEE SIS08)*, St. Louis, Missouri, USA, September 21–23, 2008.

4. D. S. Naidu, C.-H. Chen, A. Perez-Gracia, and M. P. Schoen, "Control strategies for smart prosthetic hand technology: An overview," in *Proceedings of the 30th Annual International IEEE EMBS Conference*, Vancouver, Canada, pp. 4314–4317, August 20–24, 2008 (In *Top 20 Articles, in the Domain of Article 19163667, Since its Publication (2008)* according to *BioMedLib: "Who is Publishing in My Domain?,"* Ranked as No. 8 of 20 as on August 1, 2014, Ranked as No. 9 of 20 as on May 4, 2015).

5. C.-H. Chen, D. S. Naidu, A. Perez-Gracia, and M. P. Schoen, "Fusion of hard and soft control techniques for prosthetic hand," in *Proceedings of the International Association of Science and Technology for Development (IASTED) International Conference on Intelligent Systems and Control (ISC 2008)*, Orlando, Florida, USA, pp. 120–125, November 16–18, 2008.

6. C.-H. Chen, K. W. Bosworth, and M. P. Schoen, "An adaptive particle swarm method to multiple dimensional problems," in *Proceedings of the International Association of Science and Technology for Development (IASTED) International Symposium on Computational Biology and Bioinformatics (CBB 2008)*, Orlando, Florida, USA, pp. 260–265, November 16–18, 2008.

7. C.-H. Chen and D. S. Naidu, *Intelligent Control for Smart Prosthetic Hand Technology – Phase 1-Year1-Annual*, Year 1: Annual Report, Measurement and Control Engineering Research Center (MCERC), College of Engineering, Idaho State University, Pocatello, Idaho, USA, August 5, 2008.

8. C.-H. Chen, "Hybrid control strategies for smart prosthetic hand," PhD Dissertation, Idaho State University, Pocatello, Idaho, USA, May 2009.

9. C.-H. Chen, D. S. Naidu, A. Perez-Gracia, and M. P. Schoen, "Hybrid control strategy for five-fingered smart prosthetic hand," in *The 48th IEEE Conference on Decision and Control (CDC) and 28th Chinese Control Conference (CCC)*, Shanghai, P. R. China, pp. 5102–5107, December 16–18, 2009.

10. C.-H. Chen, M. P. Schoen, and K. W. Bosworth, "A condensed hybrid optimization algorithm using enhanced continuous tabu search and particle swarm optimization," in *Proceedings of ASME 2009 Dynamic Systems and Control Conference (DSCC)*, Hollywood, California, USA, October 12–14, 2009 (No. DSCC2009–2526).

11. C.-H. Chen, D. S. Naidu, A. Perez-Gracia, and M. P. Schoen, "Hybrid optimal control strategy for smart prosthetic hand," in *Proceedings of ASME 2009 Dynamic Systems and Control Conference (DSCC)*, Hollywood, California, USA, October 12–14, 2009 (No. DSCC2009–2507).

12. C.-H. Chen, D. S. Naidu, A. Perez-Gracia, and M. P. Schoen,"Hybrid adaptive control strategy for smart prosthetic hand," in *The 31st Annual International Conference of the IEEE Engineering Medicine and Biology Society (EMBC)*, Minneapolis, Minnesota, USA, pp. 5056–5059, September 2–6, 2009.

13. C.-H. Chen, D. S. Naidu, A. Perez-Gracia, and M. P. Schoen, "Hybrid of hard control and soft computing for five-fingered prosthetic hand," in *Graduate Student Research and Creative Excellence Symposium, Idaho State University*, Pocatello, Idaho, USA, April 10, 2009 (Poster presentation).

14. C.-H. Chen, K. W. Bosworth, and M. P. Schoen, "An investigation of particle swarm optimization dynamics," in *Graduate Student Research and Creative Excellence Symposium, Idaho State University*, Pocatello, Idaho, USA, April 10, 2009 (Poster presentation).

EPILOGUE **233**

16. C.-H. Chen and D. S. Naidu, *Intelligent Control for Smart Prosthetic Hand Technology – Phase 1-Final*, Final Research Report, Measurement and Control Engineering Research Center (MCERC), College of Engineering, Idaho State University, Pocatello, Idaho, USA, August 22, 2009.

17. C.-H. Chen, D. S. Naidu, and M. P. Schoen, "An adaptive control strategy for a five-fingered prosthetic hand," in *The 14th World Scientific and Engineering Academy and Society (WSEAS) International Conference on Systems, Latest Trends on Systems (Volume II)*, Corfu Island, Greece, pp. 405–410, July 22–24, 2010.

18. C.-H. Chen and D. S. Naidu, "Optimal control strategy for two-fingered smart prosthetic hand," in *Proceedings of the International Association of Science and Technology for Development (IASTED) International Conference on Robotics and Applications (RA 2010)*, Cambridge, Massachusetts, USA, pp. 190–196, November 1–3, 2010.

19. C.-H. Chen, D. S. Naidu, and M. P. Schoen, "Adaptive control for a five-fingered prosthetic hand with unknown mass and inertia," *World Scientific and Engineering Academy and Society (WSEAS) Journal on Systems*, 10: 148–161, May 2011.

20. C.-H. Chen and D. S. Naidu, "Fusion of fuzzy logic and PD control for a five-fingered smart prosthetic hand," in *Proceedings of the 2011 IEEE International Conference on Fuzzy Systems (FUZZ–IEEE 2011)*, Taipei, Taiwan, pp. 2108–2115, June 27–30, 2011.

21. D. S. Naidu and C.-H. Chen, "Automatic control techniques for smart prosthetic hand technology: An overview," chapter 12 in *Distributed Diagnosis and Home Healthcare (D₂H₂)*, *Vol. 2*, edited by U. R. Acharya, F. Molinari, T. Tamura, D. S. Naidu, and J. Suri, American Scientific Publishers, Stevenson Ranch, California, USA, pp. 201–223, 2011.

22. C.-H. Chen and D. S. Naidu, *Intelligent Control for Smart Prosthetic Hand Technology – Phase 2-Year1-Annual*, Annual Research Report, Measurement and Control Engineering Research Center (MCERC), College of Engineering, Idaho State University, Pocatello, Idaho, USA, April 25, 2011.

23. C.-H. Chen and D. S. Naidu, "Hybrid genetic algorithm PID control for a five-fingered smart prosthetic hand," in *Proceedings of the 6th International Conference on Circuits, Systems and Signals (CSS'11)*, Vouliagmeni Beach, Athens, Greece, pp. 57–63, March 7–9, 2012.

24. C.-H. Chen and D. S. Naidu, *Intelligent Control for Smart Prosthetic Hand Technology – Phase 2-Year2-Annual*, Annual Research Report, Measurement and Control Engineering Research Center (MCERC), College of Engineering, Idaho State University, Pocatello, Idaho, USA, April 25, 2012.

25. C.-H. Chen and D. S. Naidu, *Intelligent Control for Smart Prosthetic Hand Technology – Phase 2-Year3-Annual*, Annual Research Report, Measurement and Control Engineering Research Center (MCERC), College of Engineering, Idaho State University, Pocatello, Idaho, USA, December 15, 2012.

26. C.-H. Chen and D. S. Naidu, *Intelligent Control for Smart Prosthetic Hand Technology – Phase 2-Final*, Final Research Report, Measurement and Control Engineering

Research Center (MCERC), College of Engineering, Idaho State University, Pocatello, Idaho, USA, December 19, 2012.

27. C.-H. Chen, "Homocysteine- & connexin 43-regulated mechanisms for endothelial wound healing," PhD Dissertation, Idaho State University, Pocatello, Idaho, USA, May 2013.

28. C.-H. Chen and D. S. Naidu, "Hybrid control strategies for a five-finger robotic hand," *Biomedical Signal Processing and Control*, 8: 382–390, July 2013.

29. C. Potluri, M. Anugolu, M. P. Schoen, D. S. Naidu, A. Urfer, and S. Chiu, "Hybrid fusion of linear, non-linear and spectral models for the dynamic modeling of sEMG and skeletal muscle force: An application to upper extremity amputation," *Computers in Biology and Medicine: An International Journal*, 43(11): 1815–1826, November 2013.

30. C.-H. Chen and D. S. Naidu, "A modified optimal control strategy for a five-finger robotic hand," *International Journal of Robotics and Automation Technology*, 1: 3–10, November 2014.

Acknowledgments

The research was sponsored by the U.S. Department of the Army, under award number W81XWH-10-1-0128 and administered by the U.S. Army Medical Research Acquisition Activity, 820 Chandler Street, Fort Detrick, MD 21702-5014, USA. The information does not necessarily reflect the position or the policy of the Government, and no official endorsement should be inferred. For purposes of this article, information includes news releases, articles, manuscripts, brochures, advertisements, still and motion pictures, speeches, trade association proceedings, etc.

About the Authors

Cheng-Hung Chen received his B.S. degree in Mechanical Engineering from National Chung Cheng University (CCU), Chiayi, Taiwan and M.S. degree in Power Mechanical Engineering (solid mechanics major) from National Tsing Hua University (NTHU), Hsinchu, Taiwan. Dr.[2] Chen came to the United States of America and began to study biology and chemistry at Harvard University, Cambridge, Massachusetts. He then drove 2,550 miles from Boston to Pocatello to pursue his dual Ph.D. degrees (Engineering and Applied Science and Biological Sciences), M.B.A. (marketing and management major), and B.A. in Chemistry with psychology minor at Idaho State University (ISU). As Postdoctoral Research Associate at University of Massachusetts (UMASS) Amherst, he developed a robust controller for hemodialysis patients.

Cheng-Hung Chen started his training in Earth Oriented Space Science and Technology Master's program at Technical University of Munich (TUM) and Private Pilot License (PPL) at Munich Flight Academy, Munich, Germany. He quickly expanded his knowledge in aerospace, including advanced orbital mechanics, satellite navigation, orbit dynamics and robotics, spacecraft technology, photogrammetry, (microwave) remote sensing, image processing, signal processing, and earth science system. Currently, he has worked as Applications Engineer in the USA Micro-Machining Center (MMC) of Synova Inc., a Laser MicroJet© technology company with headquarters in Duillier, Switzerland. He performs laser micro-machining tests, establishes optimum parameter set, optimizes laser processes to maximize performance, quality, and precision, and meets customer expectations in industry markets of aerospace, medical healthcare, diamond and jewelry, energy, tool manufacturing, and semiconductor.

Dr.[2] Chen is an IEEE Senior Member and his research interests in engineering are solid mechanics, control system, and aerospace, including finite element modeling, numerical modeling, composite material mechanics, static/dynamic analysis, fracture mechanics, vibration analysis, heat transfer analysis, structural testing and automatic control, PID control, optimal control, adaptive/robust control, fuzzy logic, neural network, tabu search, genetic algorithm, particle swarm optimization, robotics, spacecraft technology, and orbital mechanics. His research interest in biology is human physiology, including cardiovascular, neurocognitive and cancer diseases, microcirculation, and hemodialysis. He has written 33 international publications in engineering and biology fields and served as an editorial board member/reviewer for 10 journals in biomedical engineering.

The overall goals of his career are (1) being the CEO of a non-profit foundation in aerospace combining *academia, industry, politics, market,* and *education,* (2) exploring and harvesting new precious materials from space, (3) developing aerospace medicine (space hospital, space ambulance, space factory, space transportation, etc.) in microgravity, and (4) settling human activities into outer space.

Desineni Subbaram Naidu received B.E. degree in Electrical Engineering from Sri Venkateswara University, Tirupati, Andhra Pradesh, India, M.Tech. degree in Electrical Engineering (Control Systems Engineering) and Ph.D. degree in Electrical Engineering, from Indian Institute of Technology (IIT), Kharagpur, India.

At present, Professor Naidu is Minnesota Power Jack F. Rowe Endowed Chair and Professor of Electrical Engineering, at University of Minnesota Duluth, Minnesota, USA. Professor Naidu taught and/or conducted research at IIT; Guidance and Control Division at NASA Langley Research Center, Hampton, VA; Old Domain University, Norfolk, VA; Center of Excellence in Advanced Flight Research at United States Air Force Research Laboratory, WPAFB, OH; Center of Excellence for Ships and Ocean Structures at Norwegian University of Science and Technology, Trondheim, Norway; Measurement and Control Laboratory at Swiss Federal Institute of Technology, Zurich; Nantong University, China; the University of Western Australia in Perth; Center for Industrial and Applied Mathematics at University of South Australia in Adelaide; Center for Applied and Interdisciplinary Mathematics at East China Normal University, Shanghai, China. Professor Naidu was at Idaho State University, as Professor of Electrical Engineering, Director of School of Engineering, Associate Dean of the College of Engineering, and Director of Measurement and Control Engineering Research Center.

Professor Naidu's primary areas of teaching and research include Electrical Engineering, Electric Grid, Linear Control Systems, Digital Control Systems, Optimal Control Systems, Nonlinear Control Systems, Intelligent Control Systems, Biomedical Sciences and Engineering including Prosthetics, Robotics, Mechatronics, Guidance and Control in Aerospace Systems, Trajectory Optimization, Aerobraking for Mars Mission, Orbital Mechanics, Singular Perturbations and Time Scales (SPaTS) in Control Theory and Applications, Sensing and Control of Gas Metal Arc Welding (GMAW).

Professor Naidu received several awards and recognitions including twice the Senior Research Associateships from the National Research Council of the National Academy of Sciences, elected Fellow of the Institute of Electrical and Electronic Engineers (IEEE), elected Fellow of the World Innovation Foundation, UK, and Idaho State University (ISU) Outstanding Researcher and Distinguished Researcher. He has over 200 journal and conference publications including 8 books. He has been on the editorial boards of several journals including the IEEE Transactions on Automatic Control and Optimal Control: Applications and Methods (Wiley).

IEEE PRESS SERIES ON SYSTEMS SCIENCE AND ENGINEERING

Editor:
Mengchu Zhou, *New Jersey Institute of Technology and Tongji University*

Co-Editors:
Han-Xiong Li, *City University of Hong-Kong*
Margot Weijnen, *Delft University of Technology*

The focus of this series is to introduce the advances in theory and applications of systems science and engineering to industrial practitioners, researchers, and students. This series seeks to foster system-of-systems multidisciplinary theory and tools to satisfy the needs of the industrial and academic areas to model, analyze, design, optimize and operate increasingly complex man-made systems ranging from control systems, computer systems, discrete event systems, information systems, networked systems, production systems, robotic systems, service systems, and transportation systems to Internet, sensor networks, smart grid, social network, sustainable infrastructure, and systems biology.

1. *Reinforcement and Systemic Machine Learning for Decision Making*
   Parag Kulkarni

2. *Remote Sensing and Actuation Using Unmanned Vehicles*
   Haiyang Chao and YangQuan Chen

3. *Hybrid Control and Motion Planning of Dynamical Legged Locomotion*
   Nasser Sadati, Guy A. Dumont, Kaveh Akbari Hamed, and William A. Gruver

4. *Modern Machine Learning: Techniques and Their Applications in Cartoon Animation Research*
   Jun Yu and Dachen Tao

5. *Design of Business and Scientific Workflows: A Web Service-Oriented Approach*
   Wei Tan and MengChu Zhou

6. *Operator-based Nonlinear Control Systems: Design and Applications*
   Mingcong Deng

7. *System Design and Control Integration for Advanced Manufacturing*
   Han-Xiong Li and XinJiang Lu

8. *Sustainable Solid Waste Management: A Systems Engineering Approach*
   Ni-Bin Chang and Ana Pires

9. *Contemporary Issues in Systems Science and Engineering*
   Mengchu Zhou, Han-Xiong Li, and Margot Weijnen